Antimatter, the Ultimate Mirror

In 1928, the physicist Paul Dirac predicted the existence of antimatter in a mirror world, where the electrical charges on particles would be opposite to those of ordinary matter. This mirror world is found, fleetingly, at the quantum level, with positrons the counterpart of electrons, and antiprotons the opposite of protons. This book introduces the Lewis Carroll world of antimatter without using technical language or complex equations. The author shows how the quest for symmetry in physics slowly revealed the properties of antimatter. When large particle accelerators came on line, the antimatter debris of collisions provided new clues on its properties. This is a fast-paced and lucid account of how science fiction became fact.

GORDON FRASER works at the European Laboratory for Particle Physics, Geneva, Switzerland, where he is Editor of the *CERN Courier*, a monthly magazine covering all aspects of particle physics. He has been a Visiting Lecturer in Science Communication at several universities.

GORDON FRASER

ANTIMATTER

THE ULTIMATE MIRROR

CAMBRIDGE
UNIVERSITY PRESS

530
7la

PUBLISHED BY THE PRESS SYNDICATE OF THE UNIVERSITY OF CAMBRIDGE
The Pitt Building, Trumpington Street, Cambridge, United Kingdom

CAMBRIDGE UNIVERSITY PRESS
The Edinburgh Building, Cambridge CB2 2RU, UK http://www.cup.cam.ac.uk
40 West 20th Street, New York, NY 10011–4211, USA http://www.cup.org
10 Stamford Road, Oakleigh, Melbourne 3166, Australia
Ruiz de Alarcón 13, 28014 Madrid, Spain

First published 2000

Printed in the United Kingdom at the University Press, Cambridge

Typeface Trump Mediaeval 9.5/15pt.   System QuarkXPress™ [wv]

A catalogue record for this book is available from the British Library

Library of Congress Cataloguing in Publication data
Fraser, Gordon, 1943–
  Antimatter – the ultimate mirror / Gordon Fraser.
    p. cm.
  Includes index.
  ISBN 0 521 65252 9 (hardbound)
  1. Antimatter. I. Title.

QC173.3.F73 2000
530 21–dc21   99–043749

ISBN 0 521 65252 9 hardback

Contents

Preface

Antimatter – a name so familiar, yet at the same time so impenetrable. Adding the simple prefix *anti-* to an everyday concept immediately confounds understanding and stimulates the imagination. Excellent material for science fiction from fertile minds. One of the first was the talented Isaac Asimov, inventing robots powered by brains whose pathways used antiparticles (positrons). Then came Jack Williamson's 'contra-terrene' (CT) matter. Star Trek creator Gene Roddenberry introduced faster-than-light spaceships powered by antimatter.

This science-fiction success has endowed antimatter with incredible popular appeal. In January 1996, a modest press release from CERN, the European Laboratory for Particle Physics in Geneva, Switzerland, reported that a small experiment had synthesized the first atoms of antihydrogen, the simplest form of chemical antimatter. Amplified by the science-fiction factor, the response was incredible – within hours these few anti-atoms made prime-time TV and headlines all over the world. Not since the development of nuclear weapons had fundamental physics captured so much public imagination.

Away from the hype, this book explains why antimatter is serious science – the ultimate physics. Thanks are due to Maurice Bourquin, Frank Close, Don Cundy, John Eades, James Gillies, Maurice Jacob, Rolf Landua, Dieter Möhl, Walter Oelert, Pervez Hoodbhoy, Faheem Hussain, Alvaro de Rujula, Christine Sutton and Sam Ting who have helped me along the way. Thanks also to Simon Mitton and his team at Cambridge University Press.

I   Science fiction becomes science fact

Scientists read newspapers and watch TV like anyone else, but do not expect to learn very much about their professional interests this way. They have their own ways of keeping abreast of new developments. The advance of science is carefully documented and has its own rules and protocol. But, in January 1996, it did not happen this way. Physicists all over the world, preparing to return to their research laboratories after an end-year break, were startled to learn from mass media reports that a small experiment had made a major breakthrough. 'Scientists create the fuel of science fiction', said headlines in *The Times* of London, 'Discovery could lead to a different understanding of the Universe' claimed the *Washington Post*; 'At the door of Antimatter' – *La Liberation*, 'The Gate of the Shadow Kingdom' – *Der Spiegel*. Digesting this media hype, physicists realized the experiment had synthesized the first atoms of antihydrogen, the simplest form of chemical antimatter.

This avalanche of scientific publicity was precipitated by a four-paragraph press release from CERN, the European Laboratory for Particle Physics in Geneva, Switzerland. The response was incredible – within hours this modest story made prime-time TV and hit the front page of major newspapers all over the world. News magazines in several languages had a field day. Strangely enough, it was exactly one hundred years after Wilhelm Röntgen in Würzburg mailed a letter reporting the discovery of strange 'X-rays' that had produced photographs of the bones in his wife's hand. The impact of Röntgen's discovery had been immediate, and popular newspapers teased the public with stories of 'all revealing radiation', and women were advised to wear lead-lined clothes to protect them from prying X-ray eyes. Broadcast via the internet, the CERN press release had an even more immediate impact than Röntgen's X-rays. But how had such a story caught the public imagination while the physicists were still in the dark?

The impenetrable quantum world defies understanding, but its very impenetrability can stimulate the imagination. What governs a realm of which we can form no coherent mental picture? Of all the bizarre scientific concepts of the quantum world, antimatter has become the stock of science fiction – a key to make the impossible possible. Fictional antimatter-fuelled spacecraft shuttle through the maze of space and time. Antimatter was science fact that was adopted by science fiction, but, in January 1996, that popular fiction reverted to science fact.

ATOMIC SEX CHANGE
In 1603, the German astronomer Johann Bayer plotted the positions of about 2,000 known stars in his Uranometria celestial atlas. Today we know that even our own galaxy, the Milky Way, contains about a hundred billion stars, more than ten times the human population of the world. Astronomers estimate that the Universe contains about a hundred billion galaxies, each of which must contain about as many stars as the Milky Way, so that the Universe must contain of the order of ten thousand billion billion (10^{22}) stars, as many grains of sand as would cover a country like the UK to a depth of several centimetres.

All the matter in our world, animal, vegetable or mineral, is made from atoms. But atoms are very small: there are more of them in a cube of sugar than there are stars in the Universe. Every one of these atoms in a cube of sugar is electrically perfect, but every one of these atoms is top-heavy. If there were such a thing as stellar genetics, it would be as though its laws had contrived to make every star in the Universe male.

The motive force of atoms is electricity. Atoms are composite things, but overall are electrically neutral, their constituents carrying equal amounts of positive and negative electric charge. Such a balance could result through individual positive charges inside the atom pairing together in electrical wedlock with negative partners. This is what has happened in some distant huge stars, where ordinary atoms have been crushed by the remorseless press of gravity. But, in the atoms we know, there is no pairwise matching – the segregation of the electrical sexes is complete, each atom having a cloud of negatively charged electrons orbiting a small positively charged nucleus.

Although the atom's electric charge is thus balanced, its mass is not. More than 99.9 per cent of the mass of our world is built of positive electricity. By taking atoms to pieces, we can make both positive and negative electricity, but in our world the former is heavy, while the latter is very light, and therefore much easier to make. Is this imbalance reflected in the entire Universe, or is there a compensatory world where the atomic mass is dominated by negative electricity? In a letter to the journal *Nature* in 1898, the physicist Arthur Schuster surmised 'If there is negative electricity, why not negative gold, as yellow as our own?' For thirty years, Schuster's conjecture gathered dust.

Physicists call the equations of certain theories 'beautiful', meaning they are concise, symmetric and self-contained, free of arbitrariness. If such equations say something can happen, it usually does. One example is the famous set of equations written down by the Scottish physicist James Clerk Maxwell in 1864. In the early nineteenth century, physicists had found that a current-carrying conductor generates a magnetic field, and a moving magnet generates an electric current. Electricity and magnetism are somehow reciprocal, dual forms of each other. The exact duality became enshrined in the counterpoint of Maxwell's equations for electric and magnetic fields.

In 1927, an equation written down by another British physicist, Paul Dirac, predicted a new duality which underlined what Schuster had suggested in 1898 (but by which time almost everybody had forgotten). By Dirac's time, physicists had discovered that atoms were like miniature solar systems, with electrons orbiting around the atomic periphery, and a central atomic nucleus containing protons. Unlike the solar system, electrons carried negative electric charge and protons positive, so that the distribution of electric charge inside the atom appeared as an outer negative cloud with a small positive centre. As well as carrying opposite charge, protons are much heavier than electrons, two thousand times heavier in fact, so that the electrons' contribution to the atomic mass is very small.

Dirac's new equation was supposed to describe the electron, and did so very well. But, in addition, it said that an electron must have a counterpart particle with equal but opposite electric charge. At first Dirac thought his equation belonged to the world he was used to. The oppositely charged particle in his electron equation was the proton, suggested Dirac initially. But the symmetry of Dirac's equation is the symmetry of the Universe itself, too perfect for its result to be so outrageously flawed, with one particle two thousand times heavier than the other. Beyond our top-heavy atomic world, Dirac realized, a complementary electrical symmetry has to exist, with a new kind of genetic material for its atoms. These new particles he called 'antiparticles'. This antiparticle world would be a mirror-image of our own, with its lightweight particles being electrically positive instead of negative.

After Dirac's time, physicists went on to discover many more kinds of subatomic particle, most of which are very exotic and not found inside ordinary atoms. Although not very relevant to our everyday world, these exotic particles were once part of the tapestry of the Universe in the first fraction of a second of its existence, when the temperature was about ten billion degrees. As the Universe cooled, these unstable particles decayed away and produced the structure we now know. Synthesizing these exotic particles requires supplying enough energy to recreate these primal temperatures. According to Dirac's theory, these particles too should have antiparticles.

In a piece of electrically neutral matter made up of ordinary atomic fabric, the electrical nature of the atomic structure is latent. However if the sample is put in a strong electric field, it becomes electrically distorted, with negative charges pulled to one side, and positive pushed to the other. The whole sample becomes electrically biased. When the surrounding electric field is switched off and the tension in the electrical elastic of the atoms is relaxed, the atomic charges twang back to their equilibrium position and the sample reverts to being apparently neutral.

There is an even more fundamental electrical resilience than the structure of atoms. At the creation, 'the earth was without form, and void'. The void is the flimsiest possible fabric, but even the electrical neutrality of this primordial void was split asunder into separate particles and antiparticles by the forces unleashed in the Big Bang, the explosion that gave birth to our Universe. The primordial elastic stretched in the Big Bang is still expanding, and the particles at one end of it have evolved into the world we know. But, wherever they look, physicists see only matter composed of particles. Where are the antiparticle counterparts on the other end of the primordial elastic? Particles and antiparticles appear to have gone their separate ways. But, wherever the mirror world of antiparticles is, one day it could return. When the forces of the Big Bang are finally spent, the primordial elastic connecting particles and their antiparticles could snap back and reconstitute the Void of Genesis.

Although physicists do not know where to find antiparticles, they have learned how to make them. Soon after Dirac's realization that antiparticles had to exist, in 1932 the first such antiparticle was found – the antimatter counterpart of the electron, very light and carrying one unit of positive electric charge, and therefore called the positron. The positron is a carrier of positive electricity. As physicists became more skilful, they discovered more and more examples of antiparticles. However, these isolated antiparticles are not primordial, they are not dredged from the seabed of creation. They are synthetic, created in processes which mimic on a small scale how the Big Bang first split the electrically neutral void into particle–antiparticle pairs.

Physicists gradually learned how to tame antiparticles, first positrons, then antiprotons, and built sources which supplied them on tap. But these antiparticles last only as long as the sources are kept supplied with power. As if jealous of their monopoly, matter particles greedily attack any intruder antiparticles, annihilating them to give bursts of radiation. Antiparticles have to be carefully protected, and during their protected lifetime usually remain lone antiparticles, without any allegiance to atomic shape or form. However, antimatter should obey the laws of chemistry as well as physics. Could synthetic antiparticles be used to make material – true atoms of antimatter? Even in copious supply, antiparticles invariably annihilated with the surrounding matter before carefully chosen particles and antiparticles could be 'introduced' to each other and provide the right conditions for atomic marriage.

THE FIRST ANTIMATTER

On 12 September 1995, almost a hundred years after Schuster wrote his speculative letter to *Nature*, a German physicist called Walter Oelert looked at his computer output and realized his experiment could have manufactured about a dozen atoms of antimatter. In 1993 and 1994, he had tried to achieve his goal where others had failed. Perhaps 1995 was third time lucky.

It had been a hectic few weeks, first with the experiment trying to beat the clock and then analysing the mass of resulting information. For just 48 hours over three weeks, the experimenters had been privileged to tap the most precious piped utility in the world – antiprotons. Many physicists bid for the prized antiprotons and Oelert's team were allocated just two days. By trading beam time with other experiments, Oelert was able to make the most of this narrow slot.

With the actual experiment complete, and with all the data securely piled up in the computer, then the second phase could begin – painstakingly sifting through the mass of accumulated information. A billion antiprotons had given 300,000 signals in the experiment's computer. From these, 23,000 counts had been selected as being right for further grooming, and were being carefully analysed one by one.

After two weeks of careful work, programming the computer to take account of everything the experimental team could think of, a few counts obstinately refused to fall by the wayside. 'I felt good', said Oelert. 'I was sure they were right.' The team turned to the rest of the data, and over the ensuing weeks, a total of eleven 'gold-plated' counts turned up. Were they what physicists had been waiting for most of the twentieth century to see, or were they just some cruel trick of statistics, wisps of data accidentally blown together to form a scientific mirage?

Oelert's experiment at the world's largest scientific laboratory, CERN, was a modest project by the standards of today's Big Science. The team numbered just sixteen physicists. Elsewhere at CERN, teams of hundreds of researchers were working on experiments worth hundreds of millions of dollars. Oelert's team used salvaged equipment. 'Compared to the big experiments, ours cost almost nothing', he claims.

The big physics experiments take years to plan, design and build. Then come more years of running and analysing data. The entire working life of a university researcher can be spent in a single such experiment. In contrast, Oelert's modest proposal had been submitted for approval in October 1994 and finally given the go-ahead in February 1995. Under the code-name PS210, six months later it was up and running. Approved and completed within one year, PS210 was not even listed in that year's edition of 'Experiments at CERN', the 500-page book which was supposed to list the 136 scientific experiments then under way at the laboratory. With attention focused on the big detectors and the politics of their highly international teams, few other people at CERN had even realized that PS210 had come into being. The researchers came and went almost unnoticed.

PS210's plan did not sound very spectacular. The plan was to fire a beam of antiprotons at a fine jet of xenon gas. Antiprotons do not exist naturally on Earth. They can only be synthesized, and there are two places where they are available on tap. CERN is one, the other is Fermilab, the US particle physics laboratory on the Illinois plain near Chicago. These particles are so precious that, even when an experiment is ready to run, the antiproton supply frequently has to be shared among several customers and even then is severely rationed. A small antiproton experiment like PS210 has to remain in a state of continual alertness, like runners in their starting blocks, waiting for the gun to fire. 'Once one of the students pressed the reset rather than the start button on one of the detectors, and we missed that antiproton spill', remarked Oelert wryly. But PS210, with its jet of xenon gas, had a new idea. The scheme was to use the beam of antiprotons to make still more antiparticles. With this double layer of antiparticles, there would be more chance of providing the right conditions for getting particles and antiparticles together and synthesizing atoms of antimatter.

To a beam of subatomic particles, the atomic structure of even a solid metal target looks like a mesh of chicken wire. Most of the time the particles in the beam pass straight through. Only a tiny fraction of it 'wets' the atomic mesh. Monitoring any experiment are the 'detectors', sophisticated surveillance systems taking a precision electronic snapshot each time a particle touches the target mesh. Each of these 'events', as the physicists call them, allows physicists to reconstruct what happened when the incoming particle actually hit something. As with any surveillance system, most of the recorded data is routine. The particle physicists, the policemen of the subatomic world, carefully watch for signs of anything unusual.

The experiment's computer scans the recorded data, carefully filtering off worthless background dross in the search for valuable nuggets. As in gold prospecting, the filtering frequently leaves the pan empty and the researcher/prospectors have to return to the source for more raw material. If, after repeated attempts, the experiment still reveals nothing, the experimenters move on to other territory. After a few such unsuccessful attempts, it is tempting to abandon the experimental territory and move on. But the history of physics is littered with examples of searches which have retrodden old ground and probed deeper, finally coming up with treasure. A researcher has to have imagination, insight and lots and lots of patience.

After the computer has scoured the data, occasionally the experimenters' efforts are rewarded by the flash of a bright nugget. Even then, all that glitters is not gold. Before staking an ambitious claim, the nugget has to be carefully assayed to ensure that it is not the proverbial flash in the pan. Again, science history is full of examples of premature announcements, bold claims which have not withstood the final acid test.

In science, staking a claim means writing a paper and submitting it

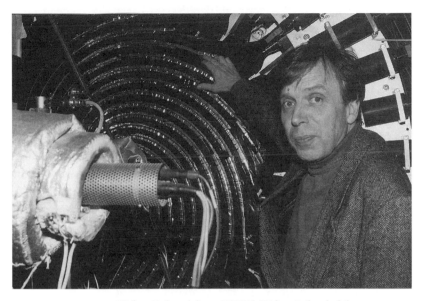

FIGURE I.I Walter Oelert (photo CERN). Walter Oelert led the team
which discovered the first atoms of chemical antimatter.

for publication in a learned journal. This 'scientific literature' is not
meant to be entertainment. Intended for other researchers, these
papers are largely incomprehensible to those not working in the field.
Even the most spectacular scientific advances are described in stilted
phrases, using obscure terminology and incomprehensible symbols.
Avoiding colourful language, the paper in time-honoured fashion sets
out what the experiment is, how it was done and, finally, what it pur-
ports to have found. The paper that Oelert's team was preparing spoke
of 'testing CPT invariance'.

PS210 had set out to make antihydrogen. Hydrogen has the simplest
of all atoms – each ordinary hydrogen atom consists of a lone electron
orbiting a single nuclear proton. Antihydrogen atoms would have a
positron orbiting a nuclear antiproton. With eleven firm antihydrogen
candidates, the PS210 team thought their dream had been realized. In
November 1995, the final draft of their paper was polished and sent to
the editorial office of *Physics Letters*, a leading European physics
research journal. Eagerly Oelert and his team awaited the outcome.

The editor of a learned journal like *Physics Letters* is chosen for his knowledge and skill in appraising research claims. However, no single person can know enough about a complicated field like particle physics to judge every paper himself. The editor normally seeks the advice of a 'referee', a knowledgeable researcher not directly involved in the experiment who can act as an impartial judge. As well as filtering off over-optimistic or charlatan papers, this refereeing process can also work in the experiment's favour, leading to suggestions to improve the presentation and quality of the result. In principle, the authors of the paper do not know who the referee is, and all correspondence goes via the editor.

For Oelert's paper, the referee was Rolf Landua, a young German researcher also working at CERN. Landua, an imaginative but careful worker who in his youth was a German champion butterfly swimmer, knew well the difficulties involved. Replying to the editor, Landua said he was not convinced that all the eleven counts were antihydrogen. Perhaps, he suggested, some of them were due to antineutrons, another form of antiparticle. Because they were electrically neutral, these anti-neutrons could be mistaken for neutral antihydrogen atoms. Anti-neutrons had been seen forty years previously. PS210's handful of nuggets should be given further scrutiny, he recommended. Realizing the anonymous referee had a good point, PS210 set to work again.

While Landua was going over the draft paper, CERN had been preparing for the December 1995 meeting of its governing body, the Council. CERN is funded by twenty European states, and biannually, in June and December, national delegates come to Geneva decide on key issues. In December, the Council traditionally has to fix the budget for the coming year. Big science is big money, CERN's annual budget being some one thousand million Swiss francs, and in cost-conscious times this budget is frequently fiercely debated and haggled to a fraction of a percentage point.

CERN's business is pure research, the furtherance of knowledge and understanding. Although in the long run this knowledge ultimately leads to technological progress, in the short term the usefulness of pure scientific progress is not easy to measure. As the *New Scientist* once said when CERN's budget was under scrutiny, the worth of such a laboratory cannot be measured in terms of non-stick frying pans or even Nobel prizes. When quizzed about the usefulness of his apparently arcane researches into electromagnetism, the nineteenth-century British physicist Michael Faraday replied 'I cannot myself imagine what use it has, but I am sure that it will one day be taxable.' Faraday's researches ultimately led to the industries of telecommunications and of electrical engineering.

Despite these difficulties in evaluating the potential of new science, in his traditional December report to the Council on the year's research achievements, CERN's Director General is naturally keen to point to concrete results and show the assembled delegates, many of whom are diplomats or civil servants rather than scientists, that they are getting value for their research investment. Christopher Llewellyn Smith, the Oxford Professor of Theoretical Physics who became CERN's Director General in January 1994, had been planning to mention the PS210 anti-matter discovery in his 1995 end-year review. Even when a significant discovery is made, the complexities of modern science make new developments difficult to explain to a lay audience. But the antimatter news was something most delegates would be able to appreciate at face value, and Llewellyn Smith had earmarked it as being speechworthy. However Landua's objections meant that any announcement of the result was premature, and Llewellyn Smith reluctantly had to stay silent on that count.

While CERN Council met, the fate of the PS210 result hung in the balance. Eleven counts surviving from 23,000 is not many, and, if most of them could be attributed to antineutrons, the experiment could not claim to have found anything. In the PS210 apparatus, the counters are segmented into three portions, each recording separately. All the data were still available for the computer, and by looking back at the way the eleven counts had been recorded in these triply segmented sensors, the experiment could tell if the signals were due to antineutrons. Carefully extending the analysis, only two of the eleven signals looked to have the characteristic antineutron signature. The PS210 team were overjoyed to find that nine bright nuggets remained. Immediately they told *Physics Letters*.

On 20 December, when most scientists were locking their laboratories and returning home for a two-week end-year break, the claims of PS210 were finally upheld and the paper was accepted for publication. The painstaking analysis had taken several months, and during this time rumours that the experiment had seen antimatter had begun to spread via the electronic grapevine of the internet. Curious scientists find it difficult to keep their mouths shut or their fingers away from their e-mail keyboards. Anxious to stop the spread of uninformed rumour, on 4 January CERN took the unusual step of issuing a press release on a scientific result before the scientific paper had been published.

CERN scientists preparing to return to work after the break were startled to hear the BBC World Service saying that antimatter had been discovered at their laboratory. CNN beamed a 64-second story worldwide. After exchanging 1996 New Year greetings and wishes, the CERN scientists eagerly sought further information. During the next few days, prime-time TV and newspaper reports piled up. The influential German weekly news magazine *Der Spiegel* ran the news as the cover story in its 15 January issue.

Walter Oelert was besieged by journalists. Arriving at Geneva for a day of newspaper interviews, he received a fax asking him to stay over until the following day so that a TV crew could also fly in. However the following day had already been reserved for another press interview, Oelert explained, this time in his home town of Jülich. After boarding the jet that evening to return to Germany, Oelert was watching the cabin crew make the final preparations for departure. Suddenly the cabin door reopened and a fax was thrust at the stewardess.

'Is there a Professor Oelert aboard?' she asked.

Oelert identified himself.

'You are asked to leave the aircraft immediately', she explained.

Oelert realized what had happened. 'I stay on this plane', he insisted.

The plane departed with Oelert on board, but it was clear that antimatter had arrived.

2 Mirror worlds

'Now, if you'll only attend, Kitty, and not talk so much. I'll tell you all my ideas about Looking-glass House', says Alice, in Lewis Carroll's *Through the Looking-Glass*. 'First there's the room you can see through the glass – that's just the same as our drawing room, only the things go the other way. I can see all of it when I get upon a chair – all but the bit just behind the fireplace. Oh! I do so wish I could see that bit! I want so much to know whether they've a fire in the winter: you never can tell, you know, unless our fire smokes, and then smoke comes up in that room too. . . . Well then, the books are something like our books, only the words go the wrong way.'

Lewis Carroll's Alice was a good scientific investigator. Observing the mirror world from her comfortable vantage-point in the real world, it was easy to assume that everything happened the same way, with a one-to-one correspondence between her own experience and what was reflected in the mirror. For everything that happened to Alice, there appeared to be a corresponding mirror-image event. But Alice was not content to take this for granted, and actually went to look. At first everything looked very familiar, but, the more she investigated, the more curious it became. There was not a one-to-one correspondence at all. When she returned from Lewis Carroll's imaginary mirror world, she had a host of new experiences to relate.

The simplest mirror, the 'looking-glass', is a sheet of flat glass with a metallic coating on one side. The image 'in' this mirror is due to light rays bouncing back from the metallic coating. But there are many kinds of mirror, real or metaphorical, each giving a recognizable image of an object. Most mirrors distort their image in some way, and it is the comparison between the object and its distorted image which provides the interest. Grossly distorting mirrors, such as those in fairgrounds, are an attraction because, despite their extreme deformation, the image is still recognizable. Mirror worlds are intriguing because they are so like our own, but nevertheless recognizably different.

Everything that happens in the real world has a counterpart in a particular mirror world, which in its own way parallels our own experience. But this image is a reflection, a transformation of the original. In a looking-glass, this transformation takes the form of a left–right reversal. Looking at the mirror, it appears as though a three-dimensional left–right inverted image 'lives' on the far side. But we know this is not true. The image in such a mirror is not real. Optics specialists say such an image is 'virtual' – it only looks as though light rays penetrate the mirror. To the eye, it appears as though light rays pass through the mirror as though it were a window and perceive a physical object on the far side. In fact the mirror is opaque, and the left–right reversed image is our brain interpreting the way light is reflected from the metallic surface.

Because what happens in the looking-glass world is in a sense the opposite of the real world, the two images, if superimposed, would somehow mutually cancel. But this cannot be done because the looking-glass image is not real. Unlike Alice, we cannot softly slide through the glass barrier. But other forms of mirror provide much more tangible reflections. Another type of reversed image is that of a black-and-white photographic negative, where light becomes dark and vice versa. The negative is a true mirror, with a one-to-one correspondence between the object and the image. However, if the positive and negative prints are superimposed, the two images mutually cancel out and all information is lost.

There are many other kinds of mirror, real and metaphorical, all relying on the stark contrast of opposites to provide depth to the reflections. Lewis Carroll's work is full of delightful contrasts, some more evident than others. In our chaotic world, what we see around us seems frequently characterized by the interplay of extremes – day and night, summer and winter. These extremes can be viewed as mirror-images of each other. However, unlike a mirror which reflects in real time, such extremes occur at different times and in different places so that they cannot actually come together. Day relentlessly follows night, and the discomfort of one sleepless night is soon forgotten. Summer relentlessly follows winter, but the time-scale is different. In Lewis Carroll's two Alice books, *Alice's Adventures in Wonderland* and *Through the Looking-Glass*, the settings are displaced by exactly six months. *Wonderland* begins on a warm May afternoon, while *Looking-Glass* finds Alice indoors on a raw day in late autumn.

Nonsense stories like Lewis Carroll's are intriguing because they are a mirror of the real world, where sense and logic can be turned upside-down and events become unfamiliar. For Alice's looking-glass, the mirror inversion turns left into right and right into left. Right and left are closely related to the idea of rotation – a rotation to the right, clockwise, reflects as an anticlockwise one. In fact the image in a looking-glass looks 'wrong', because our brain interprets it as a rotation through 180°. If we hold an object in our right hand, the mirror-image's left hand appears to be holding the object. But, if we can walk behind the mirror and turn round to face the way we came, the object is still firmly in the right hand. A clock looks very different when viewed in a mirror, but still keeps accurate time despite rotating anticlockwise. Time is unchanged by a left-to-right mirror reflection.

Electric charge, which comes in positive and negative varieties, is another example of opposite extremes. If there were such a thing as a mirror of electric charge, positive charge would be reflected as negative charge, and vice versa. With electric charge, the rule is that unlike charges attract, and like charges repel, so that in this electric mirror the image would continue to show attraction or repulsion – a positive and a negative charge would still reflect as two unlike charges, while two positive charges would appear as two negative charges, which would still repel each other.

However, there is no gadget known to science that can immediately transform positive electric charge into negative and vice versa. The amount of electric charge present always has to remain the same – like money, electric charge has to be accounted for. However, a close approximation to a mirror of electric charge is the alternating current supplied by an electrical generator. A wire loop pulled across a magnetic field generates a current, the direction of the current depending on the direction of the magnetic field and the direction of movement of

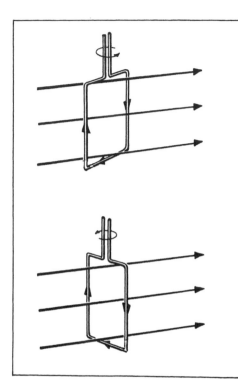

FIGURE 2.1 Electricity is sensitive to rotation. A wire loop made to rotate in a uniform magnetic field will generate an alternating current, the current changing direction as the conductor moves around the plane of the magnetic field.

the loop. Reverse the direction of the magnetic field or the direction of motion of the loop, and the current will flow in the reverse direction. The rotating loop generates an alternating current.

Electromagnetism is careful about the direction in which things happen, but electricity flowing clockwise is just as efficient as that flowing anticlockwise. In an ordinary light bulb, the current changes direction about once a second, so that the filament is continually heating up and cooling, but a video of the filament run in slow motion would reveal that the filament glows just as brightly when the current is flowing in one direction as in the other. The different phases of an alternating current provide a mirror of electric charge (Figure 2.1).

Another type of mirror is that of time, where a video of a physical process is run backwards. In most cases, such rewinds will look nonsensical, with broken objects mending themselves, dry divers emerging from splashes of water and flowers picking up dead petals. Future and past are normally very distinguishable. Once prepared, a cup of coffee always gets colder, never hotter. There is a non-zero probability that its molecules could all speed up at the same time, but countless more probabilities that the molecules will bump into each other and lose energy. In complex situations like a cup of coffee, the arrow of time is defined not by a fundamental process but by chaos constantly getting the upper hand. A child will grow up, mature and eventually die. That is certain. But how this will happen is completely unpredictable. The equations governing the child's life are too complicated to solve in advance. In such complex situations, the arrow of time has a definite direction, that of increased disorder, of delapidation.

But, in simple cases, the equations can be solved exactly. Videos of fundamental processes involving just a few interacting bodies do not show any marked change when run backwards. A video of the planets rotating around the Sun or of electrons turning around a nucleus, run in reverse, will be difficult to distinguish from one run in the correct sense. Electrons and planets do wear out. In such basic processes, the arrow of time can run equally well in either direction.

Mirror reflection, charge reversal and time reversal are examples of what mathematicians and theoretical physicists call 'transformations' – new sets of conditions imposed on Nature. In some cases, these transformations can be achieved in the laboratory, by preparing different conditions, for instance by changing the direction of a magnetic field. Other transformations are better investigated via the underlying equations. Certainly time reversal is difficult to achieve by any other means! Seeing how basic equations behave under wider and wider such transformations is the essence of modern physics and has led to great advances in our understanding.

Like Alice, physicists returning from their exploration of these mirror worlds of space reflection, of electric-charge reflection, and of time reversal, have new experiences to relate. These mirror worlds are not always like our own, and the corners where they are different can be very obscure. But they are possibly not irrelevant to our everyday experience. Electromagnetism has a 'handedness', but its mirror opposites are easy to manufacture. In fact, alternating current is more convenient to handle than direct current. Deeper in physics there are examples of handedness that cannot be reversed by changing directions in three-dimensional space. Such natural handedness, described later in this book, could have left its mark on life. All known DNA molecules, the helical strings from which living molecules are made up, are spirals which look different when viewed in a mirror. This natural prejudice has been epitomized in the title of the book by scientists John Barrow and Joseph Silk – *The Left Hand of Creation*.

THE DESTRUCTOR OF MATTER

Although it is electromagnetism that holds atoms together, we have to learn this at school. In everyday life, gravity is the most familiar natural force. But the force of gravity is so all-pervading and completely familiar that we accept it as it is. It is normal to stick to the Earth's surface and not float into space. In the relative weightlessness of outer space, astronauts can revel in being able to 'swim' around, but life quickly becomes difficult and uncomfortable. Things do not stay put. Gravity is so implanted in our consciousness as to become easily overlooked.

Which is heavier, a pound of lead or a pound of feathers? runs the old conundrum. Of course we know that a pound of anything is equivalent to a pound of anything else. Next question – which falls faster, a pound of lead, or half a pound of lead? If there were a mirror that transformed mass, how would gravity appear in this mirror? The answer is not evident at all, and was only discovered four hundred years ago. Before that, philosophers thought they knew, but had never bothered to check whether their assumption corresponded to reality.

The first to explore the mirror of mass was Galileo Galilei, born in 1564 in Pisa, Italy, the son of a minor nobleman who had fallen on hard times. Despite considerable intellectual talents, Galilei senior was only on a distant branch of his family's aristocratic tree. His relative lack of success left him embittered, and instilled in him, and probably in his children too, a contempt for unqualified authority. Galilei senior's initial plan was for his son to become a merchant, but, detecting a strong intellectual bias, he decided to send him instead to the University of Pisa. Galileo was a sceptic who took nothing for granted. Much of the scientific knowledge of those times was folklore based on presupposition, on inherited assumptions about the way things worked. Ancient philosophers were far from stupid, but in many cases were handicapped by not having adequate scientific instruments to take accurate measurements. All of the vast body of knowledge of medieval astronomy had been accumulated from observations with the naked eye. Much of medieval learning was subjective, inaccurate impressions of the world pieced together by fanciful logic, such as the fantasy that all substances were built of four basic 'elements' – earth, air, fire and water. The lack of precision measuring equipment frequently meant that many of the suppositions of this proto-science had never been tested against experiment. Nobody had looked at the effect of transformations, how systems behaved under different conditions, and, by Galileo's time, this inherited knowledge was beginning to be very creaky.

At the University of Pisa in 1582, Galileo was surprised to discover that a pendulum of a particular length always 'ticks' at the same rate, regardless of how vigorous its swing is. Most people had assumed that the swing of the pendulum would be governed by the initial jerk that set it in motion. This was the first recorded example of Galileo's healthy scientific scepticism, never taking any assumption for granted and always testing conjecture against reality. However, despite his pendulum revelation, and several others, Galileo never qualified for one of the fairly numerous scholarships for places at the University of Pisa. With his father short of money, Galileo was forced to drop out before completing the course.

Away from the university, Galileo continued to invent new instruments, and his skill and aptitude attracted more and more attention, so much so that the Duke of Tuscany recommended him as a lecturer of mathematics at the University of Pisa. In November 1589, at the age of twenty-five, and four years after having had to leave the same university in disgrace, Galileo was appointed as lecturer. Three years later, in 1592, Galileo left Pisa for the illustrious University of Padua, where he was to remain for eighteen years.

Sometime during his stay at Pisa, Galileo carried out an experiment which changed the face of scientific understanding. There is no record of the experiment ever having been carried out, but neither is there any record of Archimedes having shouted 'Eureka' from his bath. But the associated insights were certainly made. Construction of the stunning 55-metre-high white marble tower of Pisa began in 1173, but, by the time it was complete in 1350, a major design fault had become apparent. The underlying soil in the alluvial plain of the River Arno is not uniform, and, with inadequate foundations, the beautiful new tower began to lean, eventually its summit moving 4 metres out of the perpendicular. Galileo saw this architectural disaster as a custom-built laboratory for experiments on free fall. How fast did cannon-balls of different weights fall to the ground? In the fourth century BC, Aristotle had assumed that the speed of any body falling freely under gravity was proportional to its weight, and for almost two thousand years nobody thought to question this assumption, so natural did it seem. Aristotle's assumed view of the world had remained unchallenged until the sceptical Galileo appeared on the scene. Galileo took cannon-balls of various sizes to the top of the Leaning Tower and dropped them over the side. Timing how long it took for them to reach the ground, he was astonished to find that they fell at equal rates. In his *Dialogues Concerning Two New Sciences*, Galileo later wrote, 'Aristotle claims that an iron ball of one hundred pounds, falling from a height of one hundred cubits reaches the ground before a one-pound ball has fallen a single cubit. I say that they arrive at the same time. You find, on making the experiment, that the larger precedes the smaller by two finger-breadths . . . Now you would not conceal behind these two fingers the ninety-nine cubits of Aristotle.'

In subsequent epochs, such a radical discovery would have made headline news. The experiment was easily repeatable. But Galileo was aware of the stifling prejudice which surrounded the 'study' of science in those days and initially did not want to make life difficult for himself. Already he had seen that the unexpected discovery of the motion of a pendulum had not exactly helped his career. Science, resolved Galileo, had to rely on systematic experiment and demonstration, not just one example. With so much inherited prejudice, Galileo had to toil uphill all the way.

Soon Galileo moved to Padua where he was the first to use a new invention, the telescope, to observe the heavens. In another radical observation, he was the first to see that other planets had satellite moons – the Earth was not the centre of the Universe around which everything else turned. From time immemorial, everyone had assumed that the Earth had some privileged grandstand view of the heavens. No longer afraid of remaining silent, Galileo published his scientific ideas in his book *Dialogues Concerning the Two Principal Systems of the World* in 1632. It was this book which brought his unwelcome ideas to the attention of the establishment, and in 1633 the scientist was summoned to an inquisition in Rome. Refusing to budge from his heretical viewpoint, he was condemned and remained under house arrest until his death in 1642. Here he wrote his subsequent 'Dialogues' book, which had to be smuggled out of Italy before it could be published. He was only formally pardoned in 1992! Although Galileo's books were officially confiscated by Rome, they continued to circulate abroad and became required reading for inquisitive young minds.

One who had read and digested Galileo's work was a young Englishman, born in the same year that Galileo died. Isaac Newton's father died before Isaac was born, and the premature infant had been so small that he was not expected to live. However Isaac Newton went on to live to the age of 85. In 1661, he left his home in Woolsthorpe, Lincolnshire, for Cambridge University. Newton had phenomenal powers of concentration, and once his attention had been stimulated would frequently overlook sleep and meals. In 1664, he sat up one night, intrigued by a comet, and his destiny became charted. The following year, an epidemic of plague closed Cambridge University and Newton returned to his mother's house in the country. Legend has it that, in the garden of this house that autumn, Newton was thinking about celestial motion when he saw an apple fall to the ground. Eureka! Preoccupied with trying to understand celestial motion by his comet encounter, Newton realized that the falling apple, the stars and planets in their majestic orbits, and the trajectories of comets were governed by a universal force, gravity, which acts between all objects and is proportional to the amount of matter in them – their mass. The forces which act on Earth are the same which act on the heavens.

In a masterpiece of intellectual achievement, Newton conceived, developed and perfected the theory of gravity entirely on his own. In developing it, he encountered formidable mathematical obstacles which would have deterred lesser men, but the dedicated Newton derived *en route* all the new mathematics. Unaware of developments in other areas, he reinvented the mathematical edifice of differential calculus simply so that he make certain vital calculations. When later asked how he developed his monumental theory, he modestly answered 'by thinking on it'. With Newton, the verb 'to think' took on a new dimension.

In Newton's new picture, two separate quantities – mass and energy – had to balance in the detailed bookkeeping. Whatever else went on, the total mass that was present at the start had to be the same as the mass at the end – the total amount of matter in any reaction could not be changed. Likewise, the sum of all the separate energies at the beginning had to be the same as the total energies at the end. Energy could be transformed from one kind to another; for example the gravitational energy of stone could be turned into kinetic energy – the energy of motion – as it rolls down a hill, but adding up all the energies had to balance. Neither energy nor mass could disappear.

The phlegmatic Newton was pleased by his new understanding, but kept his revolutionary new ideas to himself for almost twenty years, until, in 1682, the appearance of another bright comet startled Europe and was interpreted by some as a portent of doom. But, just as the comet of 1664 had sparked Newton's interest in celestial motion, so this new

FIGURE 2.2 Isaac Newton, portrait by Thornhill, 1710 (courtesy of the Masters and Fellows of Trinity College, Cambridge). Isaac Newton realized that all masses are pulled together by the force of gravity.

comet reawakened general interest in astronomy. In London, the astronomer Edmund Halley had heard that perhaps Isaac Newton at Cambridge might be able to help him understand the new comet. Arriving at Cambridge, Halley was amazed to find that Newton had neatly formulated the laws of gravitational motion. Using Newton's theory, Halley realized that the 1682 comet was a satellite which approached the Earth every 76 years or so before retracing its path beyond the more distant planets and becoming invisible. Halley encouraged Newton to write down his theory of gravitation. Published in 1687, the famous *Philosophiæ Naturalis Principia Mathematica* became one of the most influential books ever written in science. The predicted return of Halley's comet in 1759 was a triumph for Newton's theory. The subsequent appearances of the comet were in 1835, 1910 and 1986, when it was photographed for the first time at close quarters by the Giotto space probe.

Newton's monumental picture of gravity remained unchallenged for more than 200 years. Then, in 1905, Albert Einstein, an obscure 26-year-old physicist working at the Swiss Patent Office in Berne, introduced a physics theory which went on to modify our understanding of space and time. His 'special theory of relativity' explained a puzzling paradox. Measurements had shown that the speed of light was always the same, whether the light came from a source at rest, or whether the light came from a fast-moving object like a planet. Normally an object on a moving vehicle acquires the speed of the vehicle – compared with the countryside flashing past, a man walking along a train has both the speed of the train and his own walking speed. Not so with light. The speed of light does not compound with the speed of its source. To explain this, Einstein realized that a mirror travelling at a speed comparable to that of light shows a distortion which is never seen in everyday life. The faster the mirror travels, the greater the distortion.

A 1-metre rod with a velocity v moving past an observer will appear to have a length $\sqrt{(1 - v^2/c^2)}$, where c is the velocity of light, 300,000 kilometres per second. The faster the rod moves, the shorter it appears. However, these effects only become noticeable at extremely high speeds. Under everyday conditions $v^2/c^2 = 0$. No wheeled vehicle can move fast enough for these effects to become visible, but heavenly bodies and electrons whizzing around atoms can. The mirror of relativity is real – the ultimate optical illusion – and the equations of relativity provide the transformations to translate from one mirror view to another.

Central to Einstein's relativity theory is the idea that the world is four-dimensional, the three dimensions of ordinary space being supplemented by the additional dimension of time, with information transmitted via light rays linking the space and time dimensions. No information can travel faster than a ray of light. Because its speed is not infinite, this light ray takes a certain time to travel any distance, and, when the light ray's information finally reaches its destination, it is 'old'. The light from distant stars takes many years to get to the Earth, and the image of the star we see in the sky at night is in fact an image of the star as it was many years ago. The nearest stars are a few light years away, but the light from Andromeda, the most distant object visible

FIGURE 2.3 Albert Einstein
(© Nobel Foundation). Albert
Einstein showed that matter
and energy are interchangeable,
showing the way towards
antimatter.

with the naked eye, is two million years old by the time it arrives. Mighty telescopes pick up the faint light emitted by stars several billion years ago, when the Universe was in its infancy. We have no way of knowing what these stars look like now, or even if they are there at all!

Einstein realized that the traditionally separate bookkeeping of mass and of energy could not be retained when objects move at speeds close to that of light. Moving bodies acquire energy because of their motion, but even at rest Einstein saw that each mass, m, had to be endowed with a 'rest energy' E. Mass is just another form of energy, like light or heat. Hence the famous equation $E = mc^2$. According to this equation, any expenditure of energy is accompanied by a loss in mass. But the speed of light is such a large number that the connection between energy and mass is invisible on the everyday scale. For example all the work done by a person in his or her lifetime is equivalent to a weight loss of about a tenth of a milligram – hardly a good recipe for keeping slim! (In ordinary weight loss, body fat is burnt, liberating water and carbon dioxide. These substances leave the body, which loses weight, but the total amount of mass, including the water and carbon dioxide, does not visibly change.) From an everyday viewpoint, energy – motion, heat, electricity – and mass – a measure of the amount of matter in an object – are very different. Einstein said they are equivalent and therefore interchangeable. Just as a looking-glass interchanges left and right, a relativistic mirror interchanges mass and energy.

This interchangeability is the essence of nuclear energy. Tiny amounts of mass, pawned by nuclear binding, can be liberated when nuclei are reshuffled, as in fission or fusion, releasing huge amounts of $E = mc^2$ energy. Mass adjustments amounting to a fraction of a per cent of a nuclear bomb weighing several kilograms can devastate an entire city.

Witnessing the awe of the first nuclear explosion in the Almagordo Desert in 1945, J. Robert Oppenheimer, the scientific director of the bomb project, recalled a line from the *Bhagavadgita*, the sacred book of the Hindus. Sri Krishna, the lord of fate utters : 'I am become Death, the shatterer of worlds.'

But the nuclear nightmare that Oppenheimer helped create only nibbled at the total amount of rest energy present. Is there a way of releasing all mass into energy? This is the role of antimatter, the true Destructor of Matter.

3   An imbalanced kit of electrical parts

If a crystal of sugar is crushed, a smear of the resulting powder is still sweet. 'How small can you go?' is one of the most natural questions to ask. If a piece of matter is continuously divided, is there a natural limit beyond which it cannot be cut any more while still being recognizably the same substance? The ancient Greeks realized there had to be some ultimate components of matter and built up a poetic description of the world in terms of their four classic 'elements' from which they thought all other substances could be made. Realizing the limitations of this whimsical picture, in the fourth century BC Democritos introduced the idea of atoms, from the Greek for 'uncuttable', as the ultimate division of matter. Democritos had picked up many new ideas on his travels through the Near East, and many of his viewpoints still reflect startling freshness. Some two thousand years before the invention of the astronomical telescope, he proposed that the Milky Way is a cloud of tiny stars. However, Democritus lived in the shadow of the powerful Socrates, whose followers were sceptical of other dogma. Democritus' greatest legacy was his picture of atoms of different shapes and sizes, sticking together to form substances. Because it implied that atoms had to float in an unfamiliar void, the vacuum, this practical picture did not catch on. The poetic four-element hypothesis needed no such emptiness and survived for two thousand years.

The idea of matter being composed of tiny constituent atoms lay dormant until the dawn of modern chemistry in the seventeenth century. One of the contributing intellects was Robert Boyle, the four-teenth child of the Earl of Cork, an infant prodigy who could speak fluent Greek and Latin by the age of eight. Travelling in Europe, the young Boyle studied the work of Galileo, which impressed on him the importance of empirical observations at a time when most other scientists thought otherwise. The great Dutch philosopher Spinoza tried unsuccessfully to convince Boyle that reason was better than experiment. Boyle, like his contemporary Isaac Newton, realized the importance of observation and skill in constructing apparatus and instrumentation was vital to scientific progress. Boyle constructed pumps and thermometers and was a pioneer of the chemistry of gases. These experiments convinced him that some chemical substances were more basic than others. It was these 'elements' which combined to form chemical compounds. With the publication of his famous book *Sceptical Chymist*, in 1661, the idea of the four classic Greek elements was consigned to the intellectual scrapheap.

With new apparatus available, chemistry increasingly became a quantitative science. Chemists measured how different substances interacted. In France, Joseph Gay-Lussac, Napoleon's scientific high priest, discovered in 1809 that when the gaseous elements hydrogen and oxygen combine to form water, they always do so in the same proportions by volume, no matter the scale of these volumes. This suggested that gaseous elements contained some basic units which always combine according to certain rules. In Italy, Avogadro, Count of Quaregna, interpreted these results as saying that all gases, not necessarily just the elements, contain the same number of molecules per unit volume.

In early nineteenth-century England, John Dalton had proposed a basic scenario, the 'atomic hypothesis', with all elements, not only gases, having their own distinct type of atom, and all atoms of a particular element being identical. Dalton, the son of a weaver and a gifted child, was teaching in school by the time he was twelve. Having such a youthful teacher did not impress the pupils, and Dalton turned to science instead. He proposed that each variety of atoms had its own special properties, including its 'atomic weight'. Some atoms, like hydrogen, were very light, while others were much heavier. Atoms combined together to form molecules, the smallest units of chemical compounds. In doing so, each atom had a number of links, like hooks, which could attach to other atoms. Thus a molecule of water, H_2O, contains two atoms of hydrogen and one of oxygen. Common salt, sodium chloride, NaCl, contains one atom of chlorine and one of sodium. Dalton, who was committed to the importance of weight as a means of chemical analysis, and at a time when England and France were at war, was sceptical of the results of Gay-Lussac which suggested that, for gases, volumes are more critical. Dalton also overlooked the fact that, for elements, molecules and atoms were not necessarily the same. An oxygen molecule, for example, has two atoms locked together. This oversight led to general confusion between atomic and molecular weights, which was only finally resolved by the fiery Italian chemist Stanislao Cannizzaro at the 1st International Chemical Congress, held in Karlsruhe in 1860. The atomic theory was finally ready for the textbooks.

According to this picture, each element's stock of atoms had been formed at the creation. These primeval atoms were thought to be immutable, indivisible and everlasting – the basic building material of the Universe. Atoms had survived the ruins of ancient civilizations, whose dust provided the raw materials for future generations. We are built from the same atoms as our ancestors. The air we breathe has also been breathed by every other being that has ever drawn breath.

ATOMS AND ELECTRICITY

The familiar matter formed by these atoms is all around us, but it is not easy to see that atomic matter has electrical properties. The concept of electricity, like that of atoms, dates back to the ancient Greeks. The founder of Greek science, Thales, who amazed his contemporaries by correctly predicting a solar eclipse in 585 BC, discovered that a piece of amber rubbed on his clothing could attract dust. The Greek word for amber is 'elektron', and the phenomenon became known as electricity. Electricity always had to be manufactured, the rubbing of a piece of amber on cloth being the earliest electrical device. Early electrical experiments relied on rudimentary techniques, rubbing ebonite plates to create electricity and using frogs' legs to detect it. In the eighteenth century, the technology of electricity developed rapidly, and the superindendent of gardens for the French Court, Charles-François Du Fay, discovered that electricity comes in two varieties – like kinds which repel each other and unlike kinds which attract. The Voltaic cell for the first time provided a source of electricity, and galvanometers registered electrical effects more reliably than frogs' legs.

Electricity was still a mystery, but, whatever it was, it was natural to think of it as a continuous medium, like a fluid which flowed along a conducting wire and could accumulate in a suitable material. In the mid-eighteenth century, the American statesman and scientist Benjamin Franklin proposed that an excess of this fluid was positive electricity, while a deficiency was negative electricity. An excess naturally counters a deficiency, the resultant movement explaining the flow of electricity along a conductor, said Franklin. His convention is still reflected in the way positive and negative terminals are labelled today. (In fact, electricity is usually carried by negatively charged particles, electrons, which move from the negative terminal towards the positive one, but a negative current flowing in one direction is equivalent to a positive one in the other.)

Scientists vied with each other to construct more powerful sources of electrical power. In the early nineteenth century, Humphry Davy in London built an enormous battery containing 250 metal plates. This huge power pack, built to increase the visibility of British science at a time when chemistry was dominated by French scientists, was to reveal an intimate link between chemistry and electricity. Davy, from a poor Cornish family, had become apprenticed to an apothecary, moving on to become the chemist at a fashionable clinic where gases such as nitrous oxide ('laughing gas') were used therapeutically. In 1801, Davy, a natural showman, became the lecturer at London's Royal Institution, and went on to become famous and wealthy as the discoverer of many new gases and the inventor of the miners' safety lamp.

Davy the showman was always on the lookout for spectacular experiments. Using his giant battery, he passed an electric current through molten chemical compounds. The result was a turning-point in science. Davy's demonstration experiment showed that electricity can have chemical effects. The huge current from his battery broke up compounds into their component elements, some being released at the positive electrode, others at the negative. The science of electrochemistry was born. In the space of a few years, Davy had analysed common substances like potash and common salt, discovering unexpected new chemical elements such as potassium and sodium which exploded on contact with air. Despite the fact that Britain and France were officially at war, in 1806 Davy was awarded a prize established by Napoleon for advances in electrical research. There was debate as to whether Davy should accept the honour, which he eventually did. Britain, with Davy as champion, and France, with Gay-Lussac, subsequently embarked on a scientific sovereignty race to discover new chemical elements.

However, showman Davy was more interested in displaying the spectacular effects produced by an electric current than in understanding them. The explanation of these new phenomena could only come after careful and painstaking investigation. Davy had an assistant named Michael Faraday, who would eventually carry on from where Davy left off. Born in 1791 and one of ten children of a blacksmith, Faraday's childhood was grindingly poor. At the age of fourteen, he was apprenticed to a bookbinder, where, as well as learning his new trade, he also became interested in the knowledge inside the books. In 1812, the young Faraday was given a ticket for one of the famous lectures by Davy at London's Royal Institution and was immediately captivated by his first contact with scientific experiments. A beautiful, 386-page, illustrated manuscript of Davy's lectures by Faraday impressed the vain Davy, who in 1813 took on the blacksmith's son as a factotum and bottlewasher at the Royal Institution. For this menial task, Faraday's wages were less than he had earned as an apprentice bookbinder, but science was for him far more compelling than bookbinding.

As Davy discovered more chemical elements with his electrochemical techniques, his fame spread. He was invited to embark on a circuit of prestigious European venues, the equivalent of a tour by one of today's big-name rock groups, and took Faraday as his 'roadie', in charge of all daily arrangements. Mrs Davy was demanding and cruel but Faraday stoically bore this treatment. With Davy preferring to remain a public figure, Faraday's evident scientific prowess and his skill at manipulating apparatus eventually led to him succeeding Davy as Professor at the Royal Institution in 1833, and it was in these historic premises just off London's Piccadilly that most of Faraday's electrical research was carried out.

Davy and Faraday were very different personalities. Davy was ebullient and mercurial, his research hurried and disorganized. Faraday was methodical and careful, exploring all aspects of a problem before moving on. Throughout his life, Faraday meticulously kept a personal diary, even numbering the paragraphs, eventually reaching 16,041. Faraday turned his back on the discovery of new elements and took a hard look at the curious phenomena of electrochemistry. Why did an electric current break up certain chemical compounds? Elements in their ordinary form should not carry any electric charge, but the elements released by an electric current clearly did – some, carrying negative charge, migrated to the positive electrode (called the anode by Faraday, from the Greek *anodos*, the way up), others, carrying positive charge, migrated to the negative electrode, the cathode (from the Greek *kathodos*, the way down). A true scientist, Faraday stayed with the problem, worrying about it, teasing it and making systematic measurements to test various hypotheses. He found that the amount of an element liberated was proportional to the amount of electricity passed. He also found that the quantities of elements deposited by a given amount of electricity were proportional to their atomic weights, as defined by Dalton.

In the same way that Gay-Lussac found that gaseous elements always combined in fixed amounts, Faraday's work showed that an electric current split up a compound in a definite way. It looked like Dalton's atoms were at work. In his book *Experimental Researches on Electricity*, published in 1839, Faraday wrote: 'Although we know nothing of what an atom is, we cannot resist forming some idea of a small particle, which represents it to the mind. There is an immensity of facts which justify us believing that the atoms of matter are in some

FIGURE 3.1 Michael Faraday
(Bridgeman Art Library).
Michael Faraday saw that
atoms contain electricity.

way endowed or associated with electrical powers, to which they owe their most striking qualities, and amongst them their chemical affinity.'

The particles which were liberated by an electric curent and migrated to the electrodes Faraday called ions, from the Greek verb 'to go'. What was the link between these ions and Dalton's atoms? In some cases, the substances liberated at electrodes showed that ions appeared to include combinations of elements, so that the ions had to contain several atoms, while, in others, the elements were produced and the ions were clearly more atom-like. An electric current breaks down water into its component elements, hydrogen and oxygen. However, hydrogen and oxygen atoms by themselves are electrically neutral, but the hydrogen- and oxygen-like ions liberated by an electric current somehow have their atoms distorted to carry a net electric charge. Faraday found this 'electrification' of atoms difficult to understand, and reluctantly turned to his other researches in electricity and magnetism, where he had more success.

During this time, the relationship between Faraday and his

one-time master Humphry Davy had soured. When Davy became President of the prestigious Royal Society, he tried, unsuccessfully, to block Faraday's election. In later years, Davy began to have serious health problems, probably poisoned by the chemicals he had worked with. Faraday took over the public lectures at the Royal Institution, where the humble blacksmith's son was a bigger box-office draw than his master had been. Prince Albert, the husband of Queen Victoria, and his children, were regular patrons. So was Charles Dickens. In 1844, Faraday was invited to Sunday lunch with Queen Victoria. Following his marriage, the scientist had become a member of an obscure ascetic Christian sect and was duty-bound to appear at his church every Sunday. After anguished thought, he accepted the royal invitation, but was subsequently banished by his unforgiving congregation.

Towards the end of his life, Faraday became more and more incommunicative, probably, like Davy, the victim of accumulated poisoning from a lifetime of hazardous scientific experiments where few precautions had been taken. He died in 1867. His meticulously kept diary was finally published in seven volumes in 1932. The traditional all-ticket 'Faraday' Christmas lectures at the Royal Institution, named in his memory, continue to draw large crowds of enthusiastic young students.

ELECTROCHEMISTRY

With the link between atoms and Faraday's ions still a mystery, other scientists slowly began to put together the pieces of the electrochemistry puzzle. Some substances, when dissolved in water, conduct electricity. Common salt – sodium chloride – is an example. Other substances, like sugar, dissolve but the solutions do not conduct electricity. All these substances, when dissolved in water, lower its freezing point. The molecular weight of the dissolved substance also plays a role – the lighter the molecules, the more the freezing point is lowered. Thus a gram of glucose dissolved in water freezes at a lower temperature than a gram of sucrose dissolved in water. The effect depends on the number of molecules – glucose molecules are lighter than those of sucrose so that the glucose solution contains proportionally more molecules. Common salt also lowers the freezing point, but about twice as much as expected by comparison with non-conducting solutions like glucose and sucrose.

This unexplained factor of two fascinated the gifted young Swedish scientist Svante August Arrhenius. Born in 1859, Arrhenius went on to study the new science of electrochemistry at the University of Uppsala. He realized that it was the number of particles in the electrically conducting solution that was important, and proposed that sodium chloride molecules, when dissolved in water, dissociated into two electrical particles, one carrying sodium properties and the other those of chlorine. Of course these particles could not be metallic sodium and chlorine gas, but were instead some kind of electrically charged versions with properties very different to those of normal atoms. According to Arrhenius, Faraday's ions were atoms that had either lost or gained units of electricity. In doing so they became almost unrecognizable. It was this change from element to ion that endowed chemical compounds with properties very different to those of their constituent elements – liquid water being composed of elements which are normally gases; and common salt, necessary to life, being composed of a poisonous gas and a liquid metal that explodes on contact with air.

However, this idea of electrical atoms ran up against Dalton's atomic theory, which said that atoms were the ultimate constituents of matter and therefore indivisible. While still a student, Arrhenius boldly proposed that Dalton was wrong. This did not go down very well at his doctorate examination, where his complacent examiners expected him to present conventional ideas. But, recognizing that Arrhenius was intellectually gifted, and reluctant to actually condemn a clearly interesting theory, they gave him the lowest possible pass mark. The ambitious Arrhenius sent copies of his dissertation to other chemists, and in August 1884 the German chemist Wilhelm Ostwald travelled from Riga to Uppsala to offer Arrhenius a job. Realizing they could have been mistaken in their appraisal of Arrhenius, the Uppsala authorities made a counter offer, and he was able to stay in Sweden.

Gradually Arrhenius' explanation of electrochemistry began to

FIGURE 3.2 Svante August Arrhenius (© Nobel Foundation). Svante Arrhenius won the Nobel Prize for Chemistry in 1903 for elucidating the connection between atoms and electrically charged particles, ions.

attract attention. Speaking at a Faraday lecture in London in 1881, the German physicist Hermann von Helmholtz said: 'if we accept that elementary substances are composed of atoms, we cannot avoid concluding that electricity also, positive as well as negative, is divided into definite elementary portions, which behave like atoms of electricity'. In 1874 the Irish physicist George Stoney had calculated the electric charge carried by a single such unit, calling it the 'electron'.

CATHODE RAYS, BUT NO ANODE RAYS
City centres compete for the most gaudy effects with 'neon lights' – the beautiful colours produced when electricity is passed through a tube of gas at low pressure. These effects had already been seen by Benjamin Franklin in the United States at the end of the eighteenth century, but remained, like most electrical phenomena, a curiosity until Michael Faraday in London started to explore them systematically. In 1838, he saw that the glowing column of gas in a tube was not uniform – there was a dark space next to the cathode. This, ominously called the

'Faraday Dark Space', showed that electricity had a preferred direction in the gas. At a time when electrical displays were as much entertainment as science, the far-sighted Faraday prophesied 'the results connected with the different conditions of positive and negative discharge will have a far greater influence on the philosophy of electrical science than we at present imagine'.

Julius Plücker in Bonn, Germany, found that the glow in a gas moved when a magnet was placed next to the tube. Plücker's student, Johann Hittorf, showed that these 'glow rays' came from the cathode, and, in 1876, Eugen Goldstein in Germany showed that the glow travelled in exact straight lines, casting sharp shadows of metallic objects mounted inside the discharge tube. Goldstein introduced the term 'cathode rays'. Fierce debate raged as to their origin: German scientists maintained they were radiation, like light, while their British counterparts, championed by the eccentric Sir William Crookes, said they were some kind of basic particle. The son of a poor tailor, Crookes, who went on to become one of the most influential men in Britain, remained scientifically productive throughout his life. 'We seem at last to have within our grasp and obedient to our control the little indivisible particles which with good warrant are supposed to constitute the physical basis of the Universe', wrote Crookes in 1879.

All over the world, little glass tubes, the forerunners of today's television tube, were built to investigate cathode rays. While Goldstein had shown that thick metal objects sealed inside the tubes cast shadows, thin foils let the rays through, showing they had some penetratative power. In 1895, Jean Perrin in Paris ingeniously placed a small metal cylinder inside a tube and collected the electric charge carried by the cathode rays. As would be expected for something which started from the cathode and went towards the anode, the charge was negative, but this was the first time it had been shown explicity.

In 1897, Emil Wiechert in Königsberg bent cathode rays with a magnet. Assuming that each cathode ray particle carried the charge of one Stoney electron, Wiechert estimated the mass of each of these particles to be only a few thousandths that of a hydrogen atom. For the first time, physicists realized they were dealing with subatomic particles. At Cambridge, Joseph John (J.J.) Thomson subjected his cathode ray tube to a combination of electric and magnetic fields, did not assume anything about the charge carried by the rays and made an exact measurement of the ratio of their charge to their mass.

The son of a bookseller, Thomson, like Faraday, had been exposed to books at an early age. At Cambridge, he succeeded Lord Rayleigh as Professor of Physics and Head of the new Cavendish Laboratory. Under Thomson, the Cavendish Laboratory began to specialize in the new physics of cathode rays. Thomson's work confirmed that electricity in gases was carried by negative, very light particles, and in 1899 he concluded that the cathode ray particles were Stoney's electrons, each carrying a fixed quantity of negative charge but weighing only a two-thousandth of the mass of the hydrogen atom. Negatively charged electrons were somehow embedded in atoms, but as atoms are heavy and much heavier than the electrons, the main part of an atom had to carry an equal and opposite amount of positive charge. With the subatomic electron discovered, Faraday's ions could be immediately understood as atoms that had either gained or lost electrons.

Electrons are very light and can be easily torn out of some atoms, for example even by rubbing. When atoms combine to form molecules, the atomic electrons provide the 'hooks'. However, in electrical conductors, these hooks lose their grip when an electric current is passed. The atoms which have lost negatively charged electrons have a net positive charge. They become positive ions, and are attracted to the cathode if an electric current is passed. On the other hand, the lost electrons stick to other atoms, which acquire a negative charge, becoming negative ions, migrating to the anode.

Addressing a meeting of British and French physicists at Dover in 1899, J.J. Thomson said 'Electrification essentially involves the splitting up of the atom, a part of the atom getting free and becoming detached.' This also confirmed Arrhenius' bold prophecy, made some fifteen years earlier. In 1903, Arrhenius received the Nobel Prize for Chemistry, the first Swede to receive a Nobel prize, for basically the same work which he had tried hard to defend in his Uppsala dissertation nineteen years previously. In 1906, J.J. Thomson was awarded the Nobel Prize for Physics for his discovery of the electron.

The final problem was to measure the tiny electric charge of the electron. People tried to do this by watching electrically charged water droplets through a microscope as they battled against gravity in a vertical electric field. If the field could be adjusted so that the upwards electrical force on the droplet exactly balanced the downward pull of gravity, then the electrical charge on the drop could be measured. The problem was that the drops usually evaporated before the measurement could be completed. The American physicist Robert Millikan, working in Chicago, had the idea of using oil droplets instead of water. Millikan used X-rays to knock electrons out of droplet atoms, and was able to measure the resultant charge on the ions. As different droplets could lose a different number of electrons, the electric field needed to halt their gravitational fall was not always the same, but Millikan's results clearly showed that the droplets always carried an integer multiple of some basic charge. This simple experiment earned Millikan the Nobel Prize for Physics in 1923, the first American-born physicist to make the trip to Stockholm.

THE POSITIVE HEAVYWEIGHT

The atom could be split into pieces carrying positive and negative charge, but the two pieces looked very different. If the atom is electrically neutral, but contains many light electrons, where is the balancing atomic positive charge that carries 99.95 per cent of the mass of the atom? Thomson proposed what he called a 'plum pudding' model of the atom, a dense sphere of positive charge with the tiny negative electrons embedded in it. But this did not explain how negative electrons were so easily ripped from atoms. In Japan in 1904, Hatari Nagaoka proposed a much more likely atomic model, where the electrons orbited around a central sphere of positive electricity, much like Saturn's rings.

In 1895, a young physics student named Ernest Rutherford arrived in Britain from New Zealand. In his pocket was a scholarship for further study. This scholarship was awarded to a New Zealand student only once every few years, and in Rutherford's year had initially been awarded to a young chemist. However, the chemist decided at the last minute to get married and stay in New Zealand, and the scholarship passed to the second choice, Rutherford. That year, the regulations at Cambridge University were changed and Rutherford became one of the first students from abroad to work at the Cavendish Laboratory under J.J. Thomson.

In New Zealand, Rutherford had carried out experiments on radio telegraphy. He brought his prototype transmitter to Britain and initially continued this work. At the same time, Guglielmo Marconi was carrying out his first experiments on radio telegraphy in Bologna, but, according to J.J. Thomson, Rutherford briefly held the world record for the longest distance radio telegraphy transmission. But Thomson was more interested in cathode rays than radio telegraphy and told Rutherford to follow the Cambridge party line. Unfazed, the young Rutherford dutifully obeyed and in doing so began a career which changed the course of science history. In 1898, Rutherford moved from Cambridge to McGill University, Montreal, where he made sense of radioactivity, the mysterious radiation emitted by uranium and other heavy elements. He showed that radioactivity was made up of two kinds of particles, one heavy and carrying positive charge, which he called alpha particles, and one light, carrying negative charge, which he called beta particles. For this work, Rutherford was awarded the Nobel Prize for Chemistry in 1908, one year before Marconi received his Nobel Physics prize for his pioneer work in radio telegraphy. By switching from radio telegraphy to atomic physics, Rutherford not only assured himself of a Nobel prize, but got it earlier! Shortly after Rutherford's discovery of beta particles, Walter Kaufmann in Germany showed that beta particles were the ubiquitous electrons.

From McGill, Rutherford moved to a senior position at Manchester, England. As a New Zealander he was not sensitive to the class distinctions of Edwardian England and simply looked for the best students, irrespective of their origins. He formed one of the first international research groups in physics, a trend which was to continue throughout twentieth-century physics. One member of Rutherford's Manchester team was Hans Geiger, who later became famous as the inventor of the Geiger counter for detecting radioactivity.

At Manchester, Rutherford looked at what happened when alpha particles from a radioactive source passed through a thin gold foil. This involved peering at a fluorescent screen through a microscope and counting the tiny flashes each time an alpha particle hit. To build up a picture of the alpha interactions, the fluorescent screen behind the alpha source had to be patiently scanned patch by patch with the microscope. The intense effort in counting the tiny flashes meant that the researchers had to work in pairs, one watching the screen and the other recording the score. Every few minutes, the pair exchanged roles.

They found that most of the alphas were only slightly deflected as they passed through the thin foil. In 1909, Rutherford told Ernest Marsden, a student of Geiger, to look in unexpected places to see if any alphas were severely deflected. There was no reason to believe that they should be, but Rutherford, like Faraday, was a painstaking experimenter who did not 'assume' and left nothing to chance. Marsden dutifully moved his recording screen and microscope from the far side of the target foil and mounted them on the 'wrong' side, near the radioactive source. Marsden's partner in counting the alphas was Rutherford himself.

To their amazement, they saw occasional alpha particles which appeared to have hit the foil and bounced back towards the radioactive source. Positively charged alpha particles, heavy by atomic standards – nearly ten thousand times heavier than an electron – were meeting something in the gold foil that stopped them in their tracks and sent them flying backwards. As Rutherford said, 'it was as though you had fired a 15-inch shell at a piece of tissue paper and it had bounced back and hit you!'.

It took Rutherford almost two years to solve this puzzle. He eventually realized he had discovered where the positive charge lives in the atom. Rather than being smeared everywhere, as everybody had supposed, the positive atomic charge is concentrated in one tiny but heavy grain right in the centre of the atom. This tiny grain – the atomic nucleus – carries almost all the mass of the atom. This heavy nucleus is so small that atoms, like outer space, are 99.99 per cent empty. Most of the time, the alpha particles went right through the atoms without meeting a substantial obstacle. However, if an incoming alpha did manage to hit the nucleus head-on, it was encountering something even heavier than itself, and ricocheted back.

The nucleus is so small that if an atom could be magnified to the size of a football pitch, its nucleus would only be about the same size as a marble! But this tiny nuclear marble holds the key to the atom. Far away from the nucleus are the orbital electrons, carrying almost no weight, but screening the positive electron charge of the nucleus. Each of Dalton's once indivisible atoms looked now more like a miniature solar system, with the solid nucleus at the centre and distant electron 'planets' revolving around it.

The next question was the structure of the nucleus itself. Were nuclei like Dalton's atoms – hard and immutable, but carrying positive electric charge? Where did this charge come from? Or could the nucleus itself be broken up? Again it was the genius of Rutherford which provided the answer. However, his research was slowed down by the First World War, where he had to serve as a consultant on several important committees. Apologizing for his absence at a meeting of an international committee on antisubmarine measures, he declared 'If, as I have reason to believe, I have disintegrated the nucleus of the atom, this is of far greater significance than the war.' Resuming full-time academic research, in 1919 he fired alpha particles from a radioactive source into a container filled with nitrogen. Behind the container was the faithful fluorescent screen and microscope arrangement to count the flashes due to any emerging particles.

The heavy alpha particles were normally absorbed after about 10 centimetres of gas. However, Rutherford's screen showed occasional bright flashes, showing that something more penetrating than alpha particles was coming out of the nitrogen. Rutherford concluded that

FIGURE 3.3 Ernest Rutherford (photo CERN). Ernest, Lord Rutherford, discovered the positively charged nucleus of the atom, much smaller than the atom itself and much heavier than the negatively charged electrons.

nitrogen nuclei (each carrying seven positive charges) struck head-on by alpha particles (now known to be helium nuclei, each carrying two positive charges) had changed into oxygen nuclei (each carrying eight positive charges) and hydrogen nuclei (each carrying one positive charge). These hydrogen nuclei were soon seen in experiments using different targets. The hydrogen nuclei which had been dislodged from the target nuclei were the electrical building blocks of all nuclei, and Rutherford called them 'protons'. Not only had Rutherford split the nucleus, showing how its electric charge was carried, but he had also shown that one nucleus could be transformed into another. Dalton's atoms, with nuclei deep inside them, were no longer immutable.

In the space of a hundred years, from Dalton to Rutherford, the underlying picture of the structure of matter had changed completely. The picture that emerged is very different to what first Dalton and then Thomson had supposed. In electrically neutral atoms, the negative charge is carried by light peripheral particles, the electrons, and the positive charge by particles some two thousand times heavier, the protons, residing in the nucleus. The new picture brought increased understanding, but a big question remained – why was Nature's kit of parts so electrically asymmetric? Could there be an electrical mirror-image of this picture, with heavy negatively charged nuclei and light-weight positively charged particles on the perimeter?

4 The quantum master

On 13 November 1995, a plaque in memory of Paul Adrien Maurice Dirac was unveiled in Westminster Abbey, London, alongside the memorial to the great Isaac Newton. Inscribed on Dirac's memorial is the equation that made him famous, the equation that predicted the existence of antimatter.

$$i\gamma \cdot \partial \psi = m\psi$$

The tradition is that on receiving a Nobel prize at the Royal Swedish Academy in Stockholm, the recipient gives a short lecture on some aspect of his work. On 12 December 1933, after receiving his prize, Dirac said – 'We must regard it rather as an accident that the Earth (and presumably the whole Solar System), contains a preponderance of negative electrons and positive protons. It is quite possible that for some of the stars it is the other way about, these stars being built up mainly of positrons (as anti-electrons had come to be called) and negative protons. In fact, there may be half of the stars of each kind . . . and there would be no way of distinguishing them.'

On 20 October 1984, Dirac died in Tallahassee, Florida, where he had lived and worked following his retirement as Lucasian Professor of Mathematics at Cambridge University, England, in 1969. His death warranted a modest obituary in *The Times* of London, but eleven years had to pass before the memorial was unveiled in Westminster Abbey.

TIGHT-LIPPED GENIUS
Throughout his life, the monk-like Dirac preferred the solitude of scientific contemplation. Even after his work had become acknowledged and he was showered with honours, he still shunned the limelight. As well as his enduring prediction of antimatter, P.A.M. Dirac (as he

always signed his name) became legendary for his taciturnity. Cosmologist Fred Hoyle says it was hard to get Paul Dirac to talk about any problem unless he had a perfect solution to it. Hoyle, then a research fellow at Cambridge, relates how he once telephoned Dirac to ask whether he would be prepared to give a physics seminar at Cambridge. 'Dirac made a remark that nobody else, in my opinion, would have conceived of', says Hoyle in his autobiography *Home is Where the Wind Blows* (1994). Dirac replied: 'I will put the telephone down for a minute and think, and then speak again.'

When an answer from Paul Dirac eventually did come, it was worth taking seriously. When Dirac was a physics student at Cambridge, the lecturer once challenged his class 'This is an extraordinarily simple result in the end, but why? Why should it work out like this?' One week later a quiet young Dirac came up to the lecturer and simply said 'Here you are.'

Dirac's introspective silences had roots in his austere upbringing. Dirac's father, Charles Adrien Ladislas Dirac, born in 1866 in Monthey, in the French-speaking part of the Swiss canton of Valais, was unhappy with the authoritarian atmosphere at home in the shadow of the Alps and left for England. He became a highly regarded teacher of French in Bristol, where in 1899 he married Florence Holten, the daughter of a ship's officer. Despite his bitter feelings about his own upbringing, in those late Victorian years Charles Dirac was a strict disciplinarian. In 1900, the Diracs had their first son, Reginald. Paul was born two years later, on 8 August 1902.

Despite his reaction against his own upbringing, Charles Dirac was as much a tyrant as his own father had been. French had to be the spoken language at the Dirac dinner-table in Bristol. As Mrs Dirac and Reginald could not speak the language well enough, they were banished to the kitchen, while Charles and his younger son sat together in the dining-room. Later Paul Dirac admitted that because he did not feel he could express himself very well in French, he preferred to stay silent at mealtimes. This reticence to speak became Paul Dirac's trademark. When he was awarded the Nobel prize, he was naturally allowed to invite his parents to share the joy. But Dirac invited only his mother.

In Bristol, the young Dirac had first attended the Merchant Venturer's School where his father taught French. Unusually, this school emphasized science and practical subjects rather than classics or the arts. Although Paul Dirac's best subject was mathematics, on leaving school at 16, parental pressure made him follow his brother Reginald into the Engineering College of Bristol University. Reginald had wanted to be a doctor, but was forced by his father to go into engineering. In 1924, Reginald Dirac committed suicide.

During the First World War, with many able men on active service, there was a lot of room at Bristol University. Engineering did not appeal to Paul Dirac, but from time to time engineering students had to plough through some mathematics. These mathematical problems fascinated Dirac. Then in 1919 came an event which clearly made a deep impression on him.

On 29 May of that year, there was to be a total solar eclipse, visible only from the tropics. According to Albert Einstein's theory of relativity, light has a tiny mass, and is therefore pulled by gravity. But this effect is tiny; light from a distant star should be bent by 0.87 seconds of arc (like a puff of wind making a bullet from a rifle just miss a coin-sized target at a distance of several kilometres) as it grazes past the Sun. Stars cannot be seen against the blinding glare of the Sun, so normally this effect is invisible. However, a solar eclipse would enable astronomers to check whether the image of a star was indeed deflected as its light went past the mass of the Sun. Even before the First World War ended, the famous British astronomer Arthur Eddington had persuaded the UK government to finance an expensive expedition to observe the eclipse of the Sun and look for the tiny deviation predicted by Einstein.

On 6 November 1919, at a joint meeting of the Royal Society and the Royal Astronomical Society in London, Eddington reported that, by comparing sightings of the same star before and during the eclipse, he could confirm Einstein's prediction. Light indeed has mass. The Chairman, Professor J.J. Thomson of Cambridge, said it was 'the most important result obtained in connection with the theory of gravitation since Newton's day'.

Einstein became a hero overnight, and the blaze of newspaper publicity about the theory of relativity captivated the young Dirac, who was unenthusiastically pushing ahead with his engineering studies. In 1921, he went to Cambridge for a highly competitive examination at St John's College and was offered a mathematics scholarship whose value – a mere seventy pounds a year – was ill-matched to its academic prestige. Although Dirac's father had suggested that his son should try for Cambridge, he would not supplement the modest funding offered by the scholarship. So Paul Dirac remained in Bristol. Ill-suited to the practical side of engineering and with an economic depression affecting the job market, he instead took up an offer to study mathematics free at Bristol University, a scheme designed then, as sixty years later, to keep bright young people from swelling the ranks of the unemployed.

In 1923, Dirac completed his mathematics education at Bristol, and, with a scholarship from the new Department of Scientific and Industrial Research, was finally able to go to Cambridge as a research student. His battle with engineering had taught him that laboratory research with delicate apparatus was not what he was cut out for. Instead he chose the intellectual discipline of mathematical physics. Still fascinated by the puzzles and paradoxes of Einsteinian relativity, he wanted to work at Cambridge with Ebenezer Cunningham, who had written books on the subject. But Cunningham, unsure of the depth of his understanding of these challenging new ideas, was not keen to confront gifted research students. Dirac was assigned instead to work with Ralph Fowler, who was more interested in atomic theory than relativity. Dirac was initially disappointed, but this forced encounter was the first rung on a ladder which would lead him, just ten years later, to his Nobel prize. Ralph Fowler was one of the few people in Britain at the time who realized another revolution in scientific understanding, just as far-reaching as that of relativity, was fomenting in continental Europe.

The fuss about relativity had overshadowed new progress in atomic physics. Until the early 1920s, physicists had handled atoms using everyday concepts: electrons were supposed to behave like billiard balls and light like water waves. The atom itself was viewed as a miniature planetary system, with remote electrons orbiting around a central nucleus. Increasingly, this simple picture did not work. Tiny subatomic particles like electrons did not appear to obey 'common-sense' rules. In the depths of the atom, Nature appeared to work in a different way. As if to mark the passage from one scientific century to another, Max Planck in Germany had proposed that, to understand the way it was related to temperature, radiation could not come in a continuous stream, but arrived instead as small droplets, or 'quanta', in much the same way that rainfall is ultimately composed of raindrops. Although rain falls as droplets, water supply engineers can overlook this detail. Assuming their commodity is continuous, they use fluid dynamics to design reservoirs and pumping systems. For subatomic work, the individual quantum raindrops come into play and fluid dynamics is no longer valid. Physicists had to learn how to work with Planck's quanta.

Atomic spectra are another example of quantum radiation. To explain the tracery of spectral lines emitted by hot atoms, Niels Bohr in Copenhagen had proposed a radical new picture of electrons orbiting around Rutherford's nucleus. Electrons could only orbit in certain ways, like the gears on a bicycle, jumping from one subatomic speed ratio to the next. While Bohr's quantum rules worked, physicists could see no underlying reason why subatomic physics had to be locked into a quantum strait-jacket. What laws ultimately governed this subatomic realm?

As Dirac himself said later in his classic textbook *The Principles of Quantum Mechanics* (1930), 'the classical tradition has been to consider the world to be an association of observable objects moving according to definite laws of force, so that one could form a mental picture in space and time of the whole scheme. This led to a physics whose aim was to make assumptions about the mechanism and forces connecting these observable objects, to account for their behaviour in the simplest possible way. It has become evident in recent years, however, that nature works on a different plan. Her fundamental laws do not govern the world as it appears in our mental picture in any very direct way, but instead they control a substratum of which we cannot form a mental picture without introducing irrelevancies.'

Whatever the outcome, the classical dynamics pioneered by Newton, and subsequently perfected by generations of gifted mathematicians, explained the poetry of celestial motion and could not be just abandoned. Somewhere there had to be a parallel which linked the twin descriptions of Nature, of the stars in their majestic motion through the cosmos and planets incessantly sweeping out their orbits on one hand, and the invisible hive of subatomic activity, on the other. To guide his own thinking, Dirac chose E.T. Whittaker's mathematical treatise on analytical dynamics. It turned out to be another wise choice.

Mathematical investigation can be a very lonely business. For mathematical physicists, there was no real centre for discussion at Cambridge in those days, although 'tea parties' in people's college rooms provided some focus on such arcane subjects as projective geometry. Although sometimes attending these meetings, Dirac thrived on solitude, steadily working through each problem until he solved it to his own satisfaction. Rather than picking up the new quantum jargon parrot fashion, Dirac shut himself off and tried to master its etymology and grammar.

Light, known to be composed of waves, sometimes behaved as though it were composed particles. In the photoelectric effect, discovered by Heinrich Hertz in 1887, light hitting a sensitive surface releases a tiny electric current – electrons. However, the energy of these electrons is not dependent on the brightness of the light. More light gives more electrons, but of the same energy. This could only be understood if light were composed of distinct particles, 'photons', each having a specific interaction at the photosensitive surface. Ordinary articles move in trajectories – straight lines or smooth curves, and bounce off each other. How could this be reconciled with electrons, which appeared to hop from one subatomic orbit to the next, each time releasing a photon flash?

DESIGNER EQUATIONS

A major step towards solving the particle/wave dilemma was made in 1923 by Louis de Broglie, the son of a French *député* (Member of Parliament) and who had spent the First World War assigned to a radio-telegraphy unit at the Eiffel Tower in Paris. De Broglie proposed that a particle should be accompanied by wave-like effects, and gave a designer equation which for the first time linked wave and particle behaviour.

In the aftermath of the First World War, culture was in ferment as vibrant new energies were unleashed. Schoenberg and Berg explored new avenues in 'classical' music, while American jazz was reshaping popular tunes. James Joyce and Franz Kafka created new trends in literature, discarding traditional ideas of narrative. The movies provided a new technological stage for the visual arts. In this rich ferment, unfettered new minds, mainly German-speaking, were probing the new problems of physics, discarding convention and trying radical new mathematical approaches. In 1925, the Austrian physicist Erwin Schrödinger, locked in a hotel room in the Swiss Alps with a girlfriend while his wife languished in Zurich, took de Broglie's particle/wave idea and applied it, producing another famous designer equation – the Schrödinger equation – which gave the right answers for the atomic spectra of hydrogen. For the first time, the hydrogen atom could be attacked mathematically. Solving the new equation showed that electrons could only move in orbits which fitted into the subatomic space in certain ways, like the resonances of a sound wave in an organ pipe. A new principle had arrived – 'wave mechanics'. But there was a price to pay. On the subatomic scale, reality was not the electron itself, but some fuzzy mathematical shadow of it, the 'wave function'.

In Germany, Werner Heisenberg, the same age as Dirac, had a totally different idea. In his theoretical workshop, Heisenberg played with matrices, two-dimensional arrays whose mathematics is very different to that of ordinary numbers. With ordinary numbers, A times B is equal to B times A. With matrices, the answer depends on the order in which the matrices are written. Matrix A times matrix B is only equal to matrix B times matrix A if the two matrices are very special. The bizarre mathematics of matrices seemed to model the unfamiliar behaviour of subatomic particles. Both Schrödinger's wave mechanics and Heisenberg's 'matrix mechanics' seemed to work, but few people could understand why, and nobody could reconcile the two approaches. Their apparent dissimilarity only underlined the physicists' lack of understanding.

Physicists had set foot in quantum territory and had a few limited sketch maps, but as yet had no definitive guide to this unfamiliar terrain. The gazetteer was to be Dirac. His customary preparation for intellectual gymnastics was to go for long solitary walks on Sundays 'to refresh his mind'. On one of his long walks in 1926, Dirac remembered something from Whittaker's book. Here was a possible key to the wave–matrix dilemma. At the same time, it provided a firm bridge between the classical mechanics of Newton and the mysterious new quantum picture. However, he could not recall the exact details and needed to consult the book. The university libraries were closed on Sundays and Dirac spent a sleepless night. 'My confidence gradually grew during the night', he said later.

The key equation had been written down by the French mathematician Siméon Poisson in 1809 in an elegant description of classical mechanics. These 'Poisson brackets' provided the missing link. Dirac took the quantities which Poisson and the classicists had treated as continuous variables and replaced them by abstract mathematical operators, but retained the form of the Poisson equation. Dirac, then still only a 24-year-old research student, sat down and carefully composed a paper 'The Fundamental Equations of Quantum Mechanics'. The ambitious title reflected Dirac's lofty goal, and his confidence that he had solved it. In personal relations, he was shy and diffident, but, like a superb athlete, he had total confidence in his own professional ability. In his paper, the unsatisfactory recipes of wave mechanics and matrix mechanics were replaced by a unified clear picture. Those who knew and admired the beauty of classical mechanics were especially impressed. Dirac's confidence increased, and using his new approach he embarked on ambitious new calculations in atomic physics. The results agreed with experiment, and Dirac's scientific fame was assured.

A research student has to submit a dissertation to obtain his doctorate, the next step on the academic ladder. Normally this dissertation is an attempt to solve a problem assigned to the student by his research supervisor. Such a problem should be difficult enough to keep a gifted student busy for several years, but solvable. The student has to have something to show, and it is unfair to give novice students fundamental problems which have defeated more mature researchers. Having introduced the young student to the challenges of real research, doctorate dissertations are usually archived. Only after obtaining the doctorate passport does a young researcher normally move on to 'real' research problems.

However, Dirac bypassed this initial step. Ralph Fowler, having introduced Dirac to modern quantum theory, could only stand aside and admire his student's progress, although he ensured that Dirac's work was quickly printed and published in the prestigious journal *Proceedings of the Royal Society*. Contributions could only come from members of the Royal Society, of which Fowler was a member while Dirac was not. Fortunately, Cambridge in the 1920s was far more receptive to Dirac's new physics ideas than Uppsala had been to Arrhenius in the 1880s. Instead of being locked in a university library and gathering dust, Dirac's doctorate dissertation on quantum mechanics quickly became required reading for others in the field, and provided a new focus for a subject which had become notoriously difficult to understand. He gave lectures to fellow students, including J. Robert Oppenheimer, later to lead the Manhattan Project to build the atomic bomb during the Second World War. Also attending the lectures was Ralph Fowler, the roles of teacher and pupil now reversed. There can be plenty of competition and rivalry at the research front, but Dirac met only respect and dignity. Heisenberg described Dirac's work as an 'extraordinary advance'.

On completing this opus, Dirac visited Copenhagen, where he worked with Niels Bohr, who had first proposed the idea that atomic electrons can only move in certain orbits. Because of Bohr, Copen-hagen was a mecca for European physicists, and there Dirac met Heisenberg. While Dirac clearly enjoyed this contact and the continual discussion, he still needed time and solitude for his most productive work. The famous Danish physicist Christian Møller, then a young student at Copenhagen, wrote: 'Often he [Dirac] sat alone in the inner-most room of the library in an uncomfortable position . . . He would spend a whole day in the same position, writing an entire article, slowly and without ever crossing anything out.' As with answering questions, Dirac had a very disciplined approach to writing, and would not commit himself to paper until all his thoughts were ordered in his head. After Copenhagen, Dirac went to Göttingen in Germany, where he met and impressed the German-speaking quantum pioneers.

THE SPINNING ELECTRON

In 1927, came another turning-point in Dirac's career. Still only 25, he was a now a front-rank scientist. However, most of his results paral-leled what had already been done by the German-speaking pioneers. Dirac had yet to produce something entirely new. At this stage, he returned to his first physics preoccupation, relativity. The equations for the new quantum theory did not obey relativity. They gave the right results for the hydrogen atom without it. Even so, realized Dirac, the lightweight electrons are the electrically charged particles most easily capable of moving at speeds close to that of light. Such fast-moving electrons would 'see' a very different picture to those moving slowly. A complete description of electrons should take into account the Ein-stein transformations which linked measurements made at very high velocities.

The equations of relativity and quantum mechanics were ready on the shelf and bringing them together in principle should not have posed any great problems. But relativity naturally involves quadratic equa-tions, involving the squares of quantities like energy and momentum:

$$E^2 = p^2c^2 + m^2c^4$$

where E is the energy of a particle, p its momentum, m its mass and c the velocity of light. Dirac knew that, if this equation was to represent the electron, the relativistic momentum p had to be replaced by some new mathematical operator. For a quantum mechanics equation momentum could not appear quadratically, p^2, it had to be linear, p. It was the square root of the relativistic equation that was relevant for quantum mechanics.

Every school student knows that for the simplest quadratic equation $x^2 = y$, then there are two solutions, one where x equals the square root of y, the other where x equals minus the square root of y. Likewise, taking the square root of the relativity equation gives two answers:

$$E = \pm \sqrt{(p^2c^2 + m^2c^4)}$$

This troubled Dirac, who knew that, if physics equations say that something can happen, it usually does. The quadratic equation for a ball thrown up in the air and moving under gravity usually gives two answers for when the ball reaches a given height. One is for the ball still moving upwards, the other when it is falling back to earth. Dirac wondered what the significance of the negative energy solution might be for the electron, but also had to discover the right form of the mathematical operator to represent the electron momentum.

Dirac was attempting to build a relativistic equation, and if he succeeded its consequences would have to be tested. However, the old-fashioned Schrödinger equation already gave the right answers for the hydrogen atom. Where could any new equation be tried out? The answer came from an unexpected quarter. Making the electron equations compatible with relativistic speeds was not the only obstacle. Another quantum quandary was the discovery that an electron also behaves as though it rotates around itself. The electron 'spins'. This rotation, too, obeyed new quantum rules. Just as an electron cannot sit anywhere in an atom and has assigned orbits, so the axis of a spinning electron cannot point at random. Like an electrical switch, it can only point in one of two possible directions, up or down.

When a particle which carries electric charge, like the electron,

FIGURE 4.1 Paul Dirac (© Nobel Foundation). Paul Dirac was the spiritual father of antimatter.

spins around its own axis, it behaves like a magnetic compass needle and will line up in the magnetic field. As well as the spin only being able to point up or down, another quantum puzzle was that the magnetic effect of these spinning electrons was exactly twice what was expected. Nobody could explain this mysterious factor of two. Anyone attempting to write down a complete equation for the electron would have to take account of quantum theory, of relativity, and of spin, and come up with all the right answers.

Trying to get to grips with electron spin, the Austrian physicist Wolfgang Pauli had invented toy 'spin matrices', little two-by-two arrays of numbers which he plugged into Heisenberg's matrix mechanics. These matrices acted as mathematical switches that ensured that electrons pointed either up or down. But neither the puzzling factor of two for the magnetic effect nor the relativity was there. In 1927, Dirac attended the annual Solvay physics conference in Brussels, where there was much talk of relativistic electron equations, and where Dirac

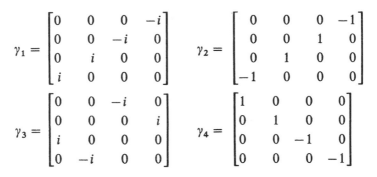

FIGURE 4.2 Dirac matrices. The four 4×4 matrices used to describe the electron. i is the square root of −1. Two rows and columns correspond to electrons, the other two to anti-electrons (positrons). The γ-notation used now has superseded the one originally introduced by Dirac. Using the γ-notation, Dirac's designer equation becomes much more symmetric (see equation at beginning of chapter).

met his intellectual hero, Albert Einstein. However, Einstein was sceptical of the fuzzy logic of quantum mechanics and Dirac was too
reserved to chat amicably with Einstein. The contact between them
was courteous and minimal. Returning to Cambridge, Dirac locked
himself away and within a few months produced another landmark
paper, this time 'The Quantum Theory of the Electron'. Again the use
of the definite article again reflected his own confidence.

The objective of combining quantum theory, spin and relativity was
transformed by Dirac into a mathematical exercise. He found that the
solution required matrices twice as big as Pauli's two-by-two arrays
(Figure 4.2), with Pauli's little two-by-two matrices embedded in them.
Dirac's new equation gave the right answer for the magnetic effect of a
spinning electron, complete with its mysterious factor of two. According to Dirac, the unexpected magnetic behaviour of the electron was a
natural consequence of relativity. Here was the testbed for the new
equation.

It was a masterpiece of intuition, and a perfect example of the power
of mathematics in modern physics, providing a concrete support for
what everyday logic finds difficult. As often happens in physics, this
mathematics had been invented some fifteen years previously by the

French mathematician Elie Cartan, but few had seen its potential for physics. Unacquainted with Cartan's work, just as Newton had been ignorant of Leibnitz' calculus, Dirac had worked out the mathematics all over again. But, wherever the new equation was used, it gave the right answer. Physicists began to grapple with the algebra of the unfamiliar new matrices.

But there was a price to pay for this success. The electron had to have four components to match the four-dimensional matrices. Two components corresponded to what was known – electrons that could spin upwards or electrons that could spin downwards. The other two electron components had negative energy. It was as Dirac had suspected when he had first looked at the relativity equations with their negative square root ambiguity. While physicists accepted the power of the new equation, they were openly critical of the negative energy impasse. Heisenberg called it 'the saddest chapter of modern physics'.

In 1929, while the debate raged about negative energy, oppositely charged particles, Dirac went to the United States, visiting universities and exploring national parks. Travel began to figure prominently in his life. In those days, international trips did not mean crowded airports and rapid flights. Instead, the days of inactivity on a transatlantic boat crossing or an equally long rail journey across the United States provided an opportunity for contemplation. For Dirac, it was an extension of his profitable custom of long Sunday walks.

In the United States, Dirac lectured at the University of Madison, Wisconsin. His fame by now was beginning to spread, but through no effort of his own. On hearing that a famous English scientist was visiting their university, a journalist from the *Wisconsin State Journal*, who wrote a humorous column under the name 'Roundy', went to see Dirac.

'So the other afternoon I knocks at the door of Dr. Dirac's office and a pleasant voice says "Come in." And I want to say here and now that this sentence was about the longest emitted by the doctor during our interview.'

'What do you like best in America?' asked the exasperated reporter after a few unsuccessful attempts to draw Dirac.

'Potatoes.'

'What is your favourite sport?'

'Chinese chess.'

'Do you go to the movies?'

'Yes.'

'When?'

'In 1920 – perhaps also 1930', replied Dirac laconically.

After leaving Madison, Dirac travelled west alone, visiting the big national parks and going on to Los Angeles. Returning to the mid-West, he lectured in Michigan and met up with Heisenberg in Chicago. Heisenberg, like Dirac, was a skilled mathematician. As a student in Munich, his mathematics efforts had been continually disturbed by the yapping of a dog that belonged to the mathematics professor. Heisenberg had been seduced away from pure mathematics by the embrace of relativity, which appeared to offer more excitement than the dry equations of conventional mathematical analysis. After that meeting in Chicago, the two physicists met again at the University of California at Berkeley, near San Francisco. Both had invitations to lecture in Japan, and decided to cross the Pacific together on a Japanese boat.

As their ship prepared to dock at Yokohama, a Japanese journalist heard that two famous scientists were aboard and went to look for them. Learning that the press was after him, Dirac hid, leaving Heisenberg to handle the journalist. Later, the two scientists were together on deck when the journalist approached Heisenberg again. 'I have searched all over the ship for Dirac but cannot find him', complained the reporter. Heisenberg then offered to answer the reporter's questions about Dirac, who just stood there, listening to Heisenberg's replies to questions about him.

After giving lectures and visiting Japanese monuments, the two split up, Heisenberg returning to Europe via China, India and the Middle East, while Dirac returned via the Trans-Siberian express. During these travels, Dirac had plenty of time to think about his negative energy electrons.

TAKING THE POSITRON PLUNGE

Negative energy had all sorts of bizarre implications. To bring a moving object to rest means taking away its energy – the brakes on a motor car convert the energy of the car's motion into friction. For a negative energy car, applying the brakes would make it accelerate, and applying the accelerator would make it slow down! The Dirac negative energy electrons were derisively called 'donkey electrons' – the harder they were pushed, the slower they went.

Dirac was tempted to identify his negative energy, positive charge solutions with the other subnuclear particle known at the time, the proton, so that his equation spanned the whole subatomic world known at the time. In 1930, Dirac suggested this in an article published in *Nature*. It was one of the few times he ventured into print with a wrong idea.

But protons do not have negative energies. Also, when an electron jumps from a negative to a positive energy state, it still has to retain its electric charge. An electron converting to a proton would somehow have to gain two units of electric charge. How? As such conundrums followed one after another from his theory, and as the objections mounted, Dirac taxed his imagination to find an explanation. The uninvited negative energies would somehow have to be pushed out of sight.

Dirac's solution was just as imaginative as his equation. Nature had an infinite number of negative energy slots, he proposed, but in our world all these would normally be filled up. This uniform filling of the negative energy slots is completely invisible to us, so no negative energy can be detected. But, if a negative energy electron can be dislodged, a slot becomes vacant. Such a vacancy would show up as a shortage of negative energy and a shortage of negative charge. A shortage of negative energy shows up as positive energy, and a shortage of negative charge as positive charge. A mirror for electric charge. The vacant slots Dirac called 'holes', a term which was to come in useful twenty years later with the invention of the transistor. With holes, he was able to explain the negative energy solutions to his equations without the unwelcome negative energies actually intruding.

The explanation can be likened to an immense underground car-park, where all the underground, or negative energy, electron parking places are full. Because it is underground, this huge car-park is invisible. However, if a car leaves one of the underground places and drives up to ground level, it becomes visible. The car-park is then left with one vacancy. As soon as this vacancy exists, one car already parked somewhere else can move into the vacant space, creating a vacancy elsewhere, and so on. The empty space 'propagates' through the ranks of parked cars. Underground car-parks were unknown in those days, and Dirac called his inexhaustible supply of negative energy states a 'sea' of electrons.

This meant that the vacuum could no longer be thought of as a void where nothing happened. In the new Dirac picture, the vacuum was in fact a bottomless pit of negative energy particles each carrying negative charge. This was difficult to accept, but, as quantum theory had only progressed by throwing conventional ideas out of the window, some adventurous physicists were eager to jettison another.

However, fitting protons into this imaginative scheme was difficult. If the electron and the proton were relativistic partners, a proton could dive into an electron hole and annihilate it. How could this be reconciled with the stability of the hydrogen atom, with its proton nucleus and single orbital electron? J. Robert Oppenheimer pointed out that Dirac's idea implied the hydrogen atom was in danger of collapsing. Protons and electrons would swallow each other up through particle–hole annihilations and stable atomic matter would cease to exist. In the electron–proton picture, the positive and negative components of atomic matter were highly unbalanced. How could an electron hole materialize as something 2,000 times heavier?

The German mathematician Hermann Weyl had a great feeling for beauty and poetry. Intuition played a major role in his mathematics. The new insights of Einstein's relativity attracted him to physics, where his eloquent book *Space, Time and Matter* (English translation 1921) helped explain the difficult new ideas to other scientists. Turning from relativity to the new quantum mechanics, Weyl's classic 1930 paper 'The Theory of Groups and Quantum Mechanics' put this revolutionary new physics on a firm mathematical foundation. In it, Weyl said 'This hypothesis leads to the essential equivalence of positive and negative electricity under all circumstances.'

Dirac got the message, and after three years of trying to squirm out of the alarming consequences, finally said what his designer equation had told him all along. In May 1931, he proposed 'A hole, if there were one, would be a new kind of particle, unknown to experimental physics, having the same mass and opposite charge to an electron. We may call such a particle an anti-electron.' Published in the *Proceedings of the Royal Society* in September of that year, it announced the birth of the modern idea of antimatter.

After having hesitated for several years before predicting the anti-electron, Dirac boldly supposed a similar duality had to hold for the other subnuclear particle known at the time, the proton. 'I think it is probable that negative protons exist, since as far as the theory is as yet definite, there is a complete and perfect symmetry between positive and negative electric charge, and if this symmetry is really fundamental in nature, it must be possible to reverse the charge of any kind of particle' – positively charged anti-electrons and negatively charged anti-protons. Here was the recipe to counterbalance the incomprehensible lopsidedness of the atom. The route to antimatter was formally declared open.

DIRAC'S LEGACY

Meanwhile Dirac began to write his masterpiece, *The Principles of Quantum Mechanics*. Here he carved out his own image of a totally new subject. It contains little scientific history, no illustrations, no references and no direct descriptions of physics experiments. The first edition of Dirac's new book was eagerly read by those who knew the importance of his work, but even so almost everybody had trouble following it. Realizing that he had been too ambitious, Dirac sat down and rewrote it completely. This second edition, published in 1935, has remained ever since the classic work for students who want to take quantum mechanics seriously. The stark pages, with their rigid lines of mathematics interspersed with careful explanation, remain exactly as Dirac wrote them. Clarity, accuracy, conciseness and compelling logic were his trademarks. Many scientists relish receiving proofs of their hastily written papers, as this gives them a chance to correct muddled thinking. The corrected papers are an editor's nightmare. Not so with Dirac. His handwriting was always practically free of errors and crossing out. On receiving proofs, Dirac would limit himself to correcting typographical errors, which is what the publishers hope for, but rarely happens. When asked about this discipline, Dirac once replied 'my mother used to say "think first, then write"'.

Thinking was what Dirac was best at and he knew it. But thinking required time, and Dirac did not waste time on other activities. Oppenheimer once offered to lend Dirac some books so that he would have something to read on his long ocean voyage from California to Japan. Dirac turned down the offer, saying that reading interfered with thought. On another occasion in California, Oppenheimer arranged a meeting between Dirac and two researchers who were trying to develop a new quantum theory of radiation. After they had explained their work, there was a lengthy silence. Finally Dirac asked, 'Where is the post office?' The researchers offered to show him if Dirac would give them his opinion of their work *en route*. 'I cannot do two things at once', he replied.

Honours now began to arrive. In 1930, he was elected a Fellow of the prestigious Royal Society. For Dirac, this meant that he could contribute articles to the journal *Proceedings of the Royal Society* without having to ask Fowler to act as his sponsor. In 1933, he shared the Nobel Physics Prize with Erwin Schrödinger, the inventor of wave mechanics. At first, Dirac did want to accept the award, fearing the publicity it would bring. But Ernest Rutherford told Dirac that refusing a Nobel prize would bring more publicity than receiving one. He followed Rutherford's advice but maintained as low a profile as possible. Under the headline 'The genius who fears women', the London *Sunday Dispatch* described him to be 'as shy as a gazelle'.

But on the academic scene he could not hide. Lucrative professorships at American universities were offered, but Dirac preferred to stay at Cambridge, knowing that in 1932 Joseph Larmor would be retiring as Lucasian Professor of Mathematics, the prestigious chair which had once been held for more than 30 years by Isaac Newton. Dirac went on to hold the Lucasian Chair for 37 years. His successor was the distinguished hydrodynamics specialist and mathematician, James Lighthill, who was succeeded in 1980 by Stephen Hawking.

The retired Dirac took up a position at Florida State University, Tallahassee. Travelling still figured prominently in his schedule until his health began to fail. Throughout his life, he was a scientists' scientist, always respected. Each time Dirac said something or went into print, it was worth listening to. But only about science. As a Nobel-prize-winning physicist who lived through the era of the atomic bomb, he would have been sure of a popular audience. Not Dirac. He never wrote on anything other than physics. Silence remained his custom throughout his life. He also scorned religion as irrational and scientifically irrelevant. Wolfgang Pauli summarized Dirac's attitude – 'There is no God, and Dirac is his prophet'.

Dirac is buried in Tallahassee, Florida. As well as the memorial in London's Westminster Abbey, there is a small garden dedicated to him in St Maurice, Valais, Switzerland, near the home which Dirac's father left in 1886.

5 Positive proof

The academic renown of the California Institute of Technology (universally abbreviated to Caltech) in Pasadena in the early 1920s was nowhere near that of the big Ivy League universities. But, with its nearby orange groves and mediterranean-style buildings, Caltech provided an attractive option for a young physicist from the east. Carl Anderson, a New Yorker, had made the trip to California at the age of 18 to begin his studies. This talented young man's mind, not yet cluttered with old ideas, was set on taking a closer look at cosmic rays, the mysterious particles from outer space which shower down on to the Earth's atmosphere.

At the end of the nineteenth century, scientists had discovered radioactivity, the invisible radiation given out by some substances, notably uranium. Like the early nineteenth century chemists who ignored the dangers of handling unknown substances, the late nineteenth and early twentieth-century scientists, such as Marie Curie, did not yet realize the dangers of this radiation, which plays havoc with the atoms it encounters, including those of their own bodies. Marie Curie died of leukaemia, the result of overexposure to radium. Ignoring what was happening inside their own bodies, early investigators saw that radioactivity knocks electrons out of the atoms of the surrounding air, making it slightly conductive. With the click of the Geiger counter still in the future, this induced conductivity was measured by a crude instrument called an electroscope, two gold leaves which, given an initial electric charge, pushed each other apart, like the pages of a newspaper in the wind. If the charged electroscope were placed in a gas which conducted electricity, the charge leaked away and the open leaves slowly collapsed. By comparing the rates of collapse in different gases, physicists could estimate the electrical conductivity. Air was always slightly conductive, which they assumed was due to the natural radio-activity of the Earth.

In 1910, a Jesuit priest called Theodor Wulf took an electroscope to the top of the Eiffel Tower in Paris and found that the air at the top was more conductive than the air at the bottom. Subsequently the Austrian physicist Victor Hess took electroscopes aloft in balloon flights, and found that the conductivity increased further with altitude. At 5,300 metres it was twice the level on the Earth's surface. Clearly the conductivity of air was not caused by radioactivity coming up from the Earth. The radioactivity came from the other direction – from outer space, and was gradually absorbed by the atmosphere. In 1925, Robert Millikan, who had won the Nobel Prize for Physics in 1923 for his classic experiment to measure the tiny charge on the electron, the first American to win a physics Nobel prize, proposed the name 'cosmic rays' for this extra-terrestrial radiation.

To find out what this mysterious cosmic radiation was, crude electroscopes were not enough. At Cambridge in the 1920s, Rutherford's laboratory had used a new tool, the cloud chamber, which revealed the individual tracks left by otherwise invisible particles. The cloud chamber was essentially a glass cylinder containing moist air and equipped with a piston. Pulling out the piston made the moist air in the chamber expand and cool, becoming supersaturated. During a brief period of sensitivity, usually about a quarter of a second, any invisible electrically charged radioactive particle flying through the chamber left a trail of electrical conductivity in its wake around which water vapour condensed, producing trails, miniature versions of those left by high-flying aircraft. Using the cloud chamber, Rutherford's laboratory had produced the first visible evidence of subnuclear transformations, when incoming particles from a radioactive source smashed into an atomic nucleus, breaking it up.

Looking out of the window at freshly fallen snow on a winter's morning, one can immediately detect where birds have hopped, the trail left by a stray cat, or whether the postman has passed. In the same way, the cloud chamber recorded the tell-tale tracks of all subatomic visitors.

POSITIVE COMING UP OR NEGATIVE COMING DOWN?

At Caltech in 1930, the 25-year-old Carl Anderson was planning to finish his research studies. For his final experiment at Caltech, he planned to use a cloud chamber to study the radiation from a radioactive source. To help analyse the complicated web of tracks left as particles passed through a cloud chamber, physicists used a magnet. If a particle carries an electric charge, it behaves like a tiny electric current and is deflected by a magnet. Cloud chambers fitted with magnets produce characteristic tracks in the form of beautiful spirals and whorls. The curvature of these tracks depends on the electric charge and the speed of the particle. The particle's charge governs the direction in which it swerves – positive particles veer one way, negative ones the other – while the speed of the particle governs how much it swerves. The faster the incoming particle, the 'stiffer' it is to turn, and the straighter its track. A slowly moving particle, on the other hand, is tightly gripped by the magnetic field and moves off in a tighter curve.

FIGURE 5.1 Carl Anderson (© Nobel Foundation). Carl Anderson was the first person to identify an antiparticle.

Another clue to the nature of the particles comes from the density of the cloud chamber tracks. A light particle, like an electron, flits rapidly through the chamber, like a small bird skipping along the surface of a layer of snow, hitting only a few atoms and leaving a sparse trail. However, a heavy particle, like a proton, continually crashes into gas atoms, producing lots of conductivity and leaving a much deeper rut. Occasionally a heavy particle like a proton is even brought to a complete stop. The source Anderson used was thorium, and at first the electron tracks he saw were very faint, because the particles moved quite fast. In desperation, Anderson added some alcohol to the water in his cloud chamber. The thicker vapour made the tracks more visible and easier to photograph!

Anderson's research supervisor was cosmic ray pioneer Millikan, who suggested that Anderson should build a chamber with the highest magnetic field then available, some 100,000 times that of the Earth's natural magnetic field. With the birth of the aviation industry in California, Caltech had recently established a department of aeronautical engineering. The 450-kilowatt generator from Caltech's Aeronautical Department wind tunnel was coaxed up to 600 kilowatts to provide the power needed for the big new electromagnet, which had to be water-cooled to absorb the heat produced by the electric current.

Looking at the curved tracks from his new chamber, Andersen was immediately struck by how many thin tracks, characteristic of electrons, seemed to curve the wrong way. Negatively charged particles like electrons coming from one direction are pushed to one side in a magnetic field. Positively charged particles coming from the same direction are pushed to the other side. The opposite curvature tracks could therefore be protons, but these heavy particles produce much thicker tracks than those Anderson was seeing. Anderson had seen a lot of electron tracks from his previous experiments and was sure that the opposite curvature tracks looked more like electrons than protons.

One explanation was that the opposite curvature tracks could be electrons moving upwards, away from the Earth, rather than downwards from the sky. In the same magnetic field, electrons moving upwards and downwards are pushed in opposite directions. Some people had already reported upwards-moving electrons, and had attributed them to recoils from cosmic rays hitting atoms of air close to the ground. But Anderson's chamber saw a lot of them, too many to attribute to recoils from below. The other possibility was that they were due to unknown electron-like particles coming down and carrying positive charge.

Millikan had left for a European lecture tour, and Anderson forwarded him eleven of his best cloud chamber photographs. These were eagerly displayed by Millikan, who claimed that the opposite curvature tracks were protons. However, European scientists with more experience of reading cloud chamber photographs did not agree, and told Millikan that the tracks were too faint for protons. They had to be due to electrons, said the Europeans.

On Millikan's return from Europe, Anderson also pushed the electron explanation, suggesting at first that the opposite curvature tracks were due to electrons moving upwards. Millikan insisted that cosmic rays came down and did not go up, so that the unexplained tracks had to be due to protons. But Anderson had made careful measurements of the track densities and became convinced this could not be so. However, Millikan's professorial seniority meant that printed papers signed by himself and Anderson were still insisting that the opposite curvature tracks were protons. Millikan saw it as an opportunity to put forward his own ideas about how cosmic rays were produced. These ideas had no room for new lightweight positively charged particles.

Protons, upward-moving electrons or downward-moving positives? The argument raged between Anderson and Millikan. To settle the question, Anderson had the idea of fitting a thin sheet of lead across the middle of his chamber. If the particles were coming downwards, they would lose energy in the lead and emerge on the far side moving slower, where they would then be bent more severely by the magnetic field. Conversely, upward-moving particles would be moving more slowly in the top half of the chamber. By comparing the curvature of the tracks in the top and bottom halves of his apparatus, Anderson would be able to tell for sure in which direction the particles were moving.

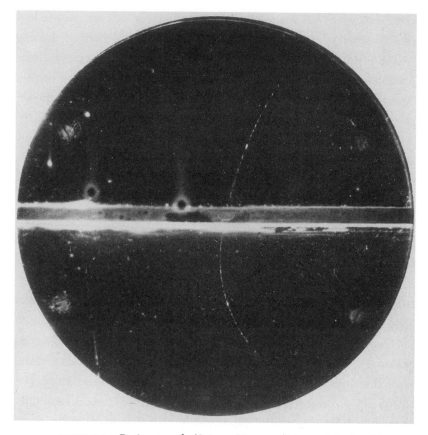

FIGURE 5.2 Positron tracks (Science Museum/Science & Society Picture
Library). The anti-electron (positron) shows up. This 1932 cloud chamber
picture shows the track of an electron-like cosmic particle passing
through a central lead plate, where it loses energy. The track curves in the
magnetic field, and the curvature increases below the plate, showing that
the particle is coming from above. The direction of curvature of the
magnetic field shows that the particle carries positive electric charge,
opposite to that of an ordinary electron. Before the idea of putting in the
lead plate, it was not clear whether such tracks were due to positive
particles coming down, or negative particles going up.

The first example found by Anderson was an electron-like particle
with positive charge moving upwards. Anderson was confused by this
double anomaly, due to a stray recoil particle from a cosmic ray hitting
an air atom somewhere below the cloud chamber and recoiling back.

However, Anderson persevered and found that almost all of the positively charged particles came down from the sky (Figure 5.2).

Meanwhile news of Anderson's findings was spreading. A photograph of one of the 'wrongly' curved tracks had been published in *Science News Letter* in December 1931. The editor of the journal convinced Anderson that the puzzling positively charged particle should be called the positron, and an option was taken on this new name. Anderson was willing to accept that the opposite curvature tracks were due to a new particle, but could not understand what it could be. He had vaguely heard about Dirac's work, but isolated at Caltech he had not understood its implications and did not know that an anti-electron had been predicted by the new theory. In September 1932, Anderson brushed aside Millikan's objections and published a note claiming that he had discovered positively charged particles whose mass 'must be small compared to the proton'. A bold step which could have wrecked his career. Fortunately he was right, but he did not make a clear link with the electron. Anderson could only suppose that the new particles somehow emerged from collisions between cosmic rays and the nuclei of atoms in the air.

EUROPEAN PAIRS

Europe was better informed about Anderson than Anderson was about Europe. In Europe, there was intense speculation about the new tracks. Was this positron the same as Dirac's anti-electron? It was tempting conclusion to jump to, but in physics every such leap has to be justified. Where did the positrons come from? According to Dirac's hole theory, an electron and a positive-energy anti-electron could be formed together when a puff of radiation penetrates the vacuum's invisible 'sea' of negative energies, knocking out an electron and leaving a hole. When created, every positron should have an electron partner.

At that time, the Cavendish Laboratory at Cambridge University under Ernest Rutherford was the world powerhouse of subnuclear physics. Under Rutherford's direction, a whole string of monumental discoveries had been or were about to be made at Cambridge by gifted students – Patrick Blackett, James Chadwick, John Cockcroft – who went on to win Nobel prizes in their own right. New techniques, such as the cloud chamber, had been developed *en route*. Even with the talent of these students, research at the Cavendish Laboratory relied heavily on the omnipresent Rutherford.

Although they were close neighbours, there was little contact between Rutherford's laboratory and Dirac's world of mathematical formulae and esoteric ideas. Rutherford saw his experiments as pointing the way forward, and whatever they discovered then had to be explained by the theorists. If there were ideas to be tested out, they were Rutherford's, worked out in homely equations, not the abstract mathematics of Dirac. At the Solvay physics conference in Brussels in 1933, where the positron discovery was discussed, Rutherford said: 'It seems to be to a certain degree regrettable that we had a theory of the positive electron before the experiments . . . I would be more pleased if the theory had appeared after the establishment of the experimental facts'! Rutherford, by then 62, was in the twilight of his career. Once his blunt logic had been able to make sense of the atom, but no more. The baton of physics prediction had passed to a new generation of mathematical specialists.

But the Rutherford school still knew better than anybody else how to do physics experiments. One student was Patrick Blackett, who had been groomed for a career as a naval officer and had served as a gunnery officer in the First World War, seeing action at the naval battle of Jutland. In 1919, four hundred British naval officers were sent to Cambridge for six months of general studies. Lieutenant Blackett liked it so much that he resigned his commission and began a new career in physics. Blackett had used the cloud chamber to record the first photograph of the breakup of an atomic nucleus. Exposed to natural radioactivity, Blackett's cloud chamber piston operated incessantly over long periods, producing many thousands of chamber expansions. Most of the resulting photographs showed nothing. As today's physicists would say, there were few 'events', when the radioactivity actually made something happen inside the chamber. To find 'events', the cloud chamber photographs had to be laboriously but vigilantly scanned. With cosmic ray tracks showing up in only a few per cent of all chamber exposures, finding photographs of subnuclear collisions was hard work.

In 1931, the Italian physicist Giuseppe ('Beppo') Occhialini came to work at the Cavendish Laboratory. Occhialini was a specialist in Geiger counters. Like the old electroscope, these counters relied on radioactivity making a gas slightly conducting. However, in the Geiger counter, equipped with high-voltage electrodes, the conducting gas produced a pulse of electricity. These pulses, heard as a characteristic series of clicks, accurately 'counted' the radioactivity. Blackett was initially sceptical about the value of these new-fangled devices. 'To make it work you had to spit on the wire on a Friday evening in Lent', he scathingly remarked. This scepticism was soon proved wrong. Geiger-counter specialists had developed an electronic circuit which could link two counters, placed one on top of the other, and register the passage of a single particle through both chambers. Using this 'coincidence circuit', and sandwiching a cloud chamber between two counters, the cloud chamber could be triggered to operate only if the two counters sparked immediately one after the other, meaning that a particle had passed through both of them, and therefore the intervening cloud chamber as well. In this way, the cloud chamber photographs became much more selective. In the early summer of 1932, 80 per cent of the photographs showed cosmic ray tracks. Blackett and Occhialini were so overjoyed at their success that at first they did not notice a few of their recorded electron tracks were bent the wrong way in the magnetic field of the cloud chamber.

Blackett talked to Dirac about these tracks, but Dirac did not push his prediction of anti-electrons and Blackett did not take Dirac's theory seriously. Only on hearing about Anderson's discovery did Blackett and Occhialini realize they had a 'great abundance' of positrons. They also saw something which Anderson, remote from Dirac's theory, had not thought of looking for – V-shaped pairs of tracks, curling off in opposite directions. These were electron–positron pairs, produced when a photon of radiation dislodged an electron from the negative-energy sea, supplying enough energy for it to become a positive-energy particle, and leaving a hole, the positron. Occhialini, less inhibited than his British colleagues, immediately rushed across to Rutherford's house with the news, kissing the astonished maid when she opened the door!

By the late autumn of 1932, Occhialini and Blackett had collected some 700 cosmic ray cloud chamber photographs. The pair of scientists were also impressed by the number of particles produced in their cosmic ray collisions, about twenty at a time, half positive charged, half negative, diverging from a common collision point. Measuring the track density and range, the scientists deduced that the masses of the positively charged particles were not much different from that of the negative electron. With equal numbers of positively and negatively charged electrons being produced, and knowing that the former do not normally exist on Earth, Blackett and Occhialini deduced that the positive–negative electron pairs were produced by otherwise invisible high-energy cosmic radiation hitting the nuclei of atoms in their cloud chamber. Each quantum of radiation produced a number of electron–positron pairs. From Einstein's equation $E = mc^2$, the energy required to produce such a pair is twice the mass of an electron (or positron). So was demonstrated for the first time the transformation of radiation into matter.

The paper was received at the London office of the *Proceedings of the Royal Society* on 7 February 1933. Meanwhile, Anderson had learnt that others were on the positive electron path, and hurried to write a conclusive paper. Anderson's article 'The Positive Electron' reached the *Physical Review* in New York on 28 February. Blackett and Occhialini had been helped by having Dirac nearby. Anderson had lost time convincing Millikan. Fortunately he had covered himself by going ahead and publishing his tentative suggestion in 1932. Robert Millikan eventually accepted the positron idea and went on to propose, controversially, that positrons were a major component of cosmic rays. In Millikan's 300-page autobiography, there are few references to Carl Anderson.

FIGURE 5.3 Patrick Blackett (photo CERN). Blackett, with Giuseppe Occhialini, was the first to see pairs of electrons and positrons being produced together.

In 1934 in Paris, Frederic Joliot and Irène Curie (the daughter of Marie and Pierre Curie) discovered new radioactive materials which emitted single positrons. Antiparticles did not only come down from the sky. Positrons are also formed in the Sun and other stars, whose thermonuclear furnaces are initially kindled by a rare nuclear transformation. Protons bounce around for thousands of millions of years, eventually avoiding the electrical repulsion of another proton to fuse with it, forming a heavier nucleus carrying a single positive charge, and spitting out a positron.

Later, Anderson remarked: 'a sagacious person . . . had he been working in a well equipped laboratory, and had he taken the Dirac theory at face value, could have discovered the positron in a single afternoon'. But the Dirac theory carried so many novel ideas with it that experimenters had not had time to adjust their thinking. Theorists, on the other hand, were too busy trying to assimilate the new ideas to suggest the right experiments to their colleagues. It is ironic that Dirac, the father of the positron, worked almost next door to the leading subnuclear laboratory in the world, and still the positron was discovered in Calfornia.

In 1936, Victor Hess, aged 53, and Carl Anderson, aged only 31, shared the Nobel Prize for Physics, the former for his pioneer work on cosmic rays, the latter for his discovery of the positron in them. Blackett won the Nobel Prize for Physics in 1948 for his lifetime's work with cloud chambers.

After contributing to many more physics discoveries, 'Beppo' Occhialini left particle physics in 1960 and turned his attention to physics experiments for space. He died in 1993. The Italian–Dutch BeppoSAX satellite (Satellite per Astronomia X), named in his honour, was launched in 1996 to study cosmic X-rays above the blanket of the Earth's atmosphere. For thirty years, astronomers had been puzzled by 'gamma-ray bursters' – intense blasts of high-energy radiation coming from anywhere in space but lasting only a few seconds. In 1997, BeppoSAX was able to see the X-rays from such a burst, localizing the explosion to a hitherto unseen X-ray star in the constellation of Orion. For the first time, one of the mysterious gamma-ray bursters was localized to a star.

6 The back passage of time

'What I am really trying to do is to bring birth to clarity, which is a half-assedly thought out pictorial semi-vision kind of thing. I would see the jiggle-jiggle-jiggle or the wiggle of the path. Even now when I talk about the influence functional, I see the coupling and I take this turn – like as if there was a big bag of stuff – and try and collect it and push it'. Thus Richard Feynman, mathematics virtuoso, physics genius, safecracker, rogue, bongo drummer and Nobel prizewinner, explained how he intuitively followed the behaviour of an electron. Feynman's new picture of the electron, taking account of what could happen in the future as well as what had happened in the past, provided a natural setting for the positron. No longer did physicists have to think in terms of abstract vacancies and holes.

For Feynman's jiggle-wiggle intuition to get to work, the ground had been carefully cultivated by previous generations of meticulous but less flamboyant physicists. Maxwell's classic equations had shown that electromagnetism is a vibrating wave which travels at the speed of light – electromagnetic radiation. Understanding the electron first meant understanding this radiation, as radiation is the messenger of electrons and all other charged particles. Without radiation, electric charge has no purpose, an airport with no aircraft.

ATOMIC OSCILLATORS

Towards the end of the nineteenth century, physicists had been trying to explain how hot objects give out electromagnetic radiation in the form of heat and light. As an object is heated, its atomic oscillators move faster, acting as tiny dipole aerials emitting electromagnetic radiation. First the object simply gets hot, then it glows dull red, then orange, and finally becomes 'white hot'. Physicists found that the radiation given out by hot material depends only on temperature, and not on the material itself. In trying to understand the underlying mechanism of atomic oscillators, physicists had to be able to predict the spectrum of this radiation – how it is made up of radiation of different wavelengths, how much heat, how much light, how much infra-red, how much ultra-violet. As the temperature increases, shorter wavelengths come into play – light has a shorter wavelength than heat. The physicists' model of atomic oscillators gave a spectrum which corresponded to what was seen, except in one place. The equations clearly said that very hot bodies would release an infinite amount of small wavelength radiation – ultra-violet light. This was clearly wrong. Nothing can be an inexhaustible source of radiation.

When the equations of physics give firm predictions, such as Dirac's suggestion of the negative square root, these predictions are worth following. But when the equations throw up unwanted infinities, it usually means that they have reached the end of their shelf life and the basic assumptions have to be rethought. Although the equations are strictly 'wrong', it does not mean that they have to be discarded. As long as they work well in one particular domain, they are useful. Warm and dry inside, one can ignore the rain outside.

In 1900, Max Planck in Berlin was looking for a way to avoid the ultra-violet downpours predicted by the new radiation equations. To force the frequency profile of the radiation into the form seen by experiments, Planck proposed that the vibration energy E of the atomic oscillators, instead of being equally shared among all oscillating frequencies, is allotted according to the frequency, v: $E = hv$, where h was some special number, Planck's constant. As frequency is inversely proportional to wavelength, this relation can also be written $E = hc/\lambda$ where c is the velocity of light and λ the wavelength of the radiation.

The new recipe appeared to work. The high temperature infinities disappeared, and the frequency spectrum followed closely what the experiments saw. But in smoothing the frequency spectrum, Planck had opened a new physics door. Putting $E = hv$ implied that radiation is not a continuous stream, like a river, but came instead as raindrops.

The atomic oscillators emit radiation as a series of pulses, like a flashing light. Plank termed each of these pulses a 'quantum' (plural quanta) from the Latin 'quantus' – how much? The size of the quanta are inversely proportional to the wavelength of the radiation. The longer the wavelength, the smaller and less noticeable the quanta, so that long wavelength radiation, such as radio waves, can be approximated to a steady stream. However, as the wavelength gets smaller, the raindrops have more energy, and ultra-violet radiation appears more 'grainy'. At the time, Planck did not really believe in these quanta as real physical entities. They were, he supposed, a mathematical trick which ensured that everything worked out right.

This quantization of radiation is like the packaging of liquids. Mineral water is sold in large bottles. Other liquids, like perfume, are more expensive and are sold in small phials. Planck's new recipe also provided unexpected bonuses. Some substances can give off electrons – a tiny electric current – when exposed to light. If light were a continuous stream, more light would simply mean more electrons. However, in this 'photoelectric effect', the energy of the liberated electrons depends more on the wavelength of the light than its intensity. To the photoelectric substance, the light looks so grainy that it is better to talk of the individual light quanta as particles – 'photons'. Each light photon pushes one electron out of the photosensitive material, the energy of the electron depending on the energy of the photon.

As quantum theory developed in the early twentieth century, these quanta of radiation became understood because electrons in atoms appear to behave like elevators in buildings – they only stop at the floors and not in between. An electron moving from one energy floor to another either emits or absorbs a definite quantum of radiation, depending on the distance between the energy floors and whether it moves to a higher or lower energy floor. These fixed energies, and therefore wavelengths according to Planck's equation, show up as spectral lines, the characteristic 'fingerprint' of light from different atoms when passed through a prism. This picture was perfected in the mid-1920s when the new Schrödinger/Heisenberg/Dirac quantum mechanics enabled the discrete energy levels of electrons in atoms to be calculated.

Physicists now understood why radiation was emitted in quantum packets, and the triumphal Dirac equation provided an almost perfect description of the electron. But Dirac could not explain why electric charge could only exist in multiples of a fixed unit, the electronic charge. Dirac's picture was a static one, saying how electrons had to coexist peacefully with the nucleus. The electron is the electrically charged oscillator responsible for electromagnetic radiation. The Dirac equation did not say what would happen if an electron was given a sudden impulse. How much radiation would it emit? If the electron were irradiated, how much would it absorb? Despite the new understanding of the electron as a particle, physicists understood very little about it as an oscillator. The electron was to quantum mechanics what Planck's light quantum had been to Maxwell's electrodynamics, commented Heisenberg in 1929.

The first attempt at describing the interaction of electrons and radiation – quantum electrodynamics – had been made by Pascual Jordan in Germany in 1926. When faced with a confusing situation, a favourite physics approach is first to make as many simplifying assumptions as possible. With this 'toy' model, physicists can test new ideas one by one. If the model does not work, it is simply discarded. If it works, it can be refined. Jordan considered matter as an array of charged particles which oscillated when irradiated, like springs being shaken. These twanging electric springs produced new vibrations, which Jordan worked out by assuming that these vibrations were themselves like a string, fixed at both ends, and could therefore only vibrate at certain frequencies. It was the Wright biplane version of quantum electrodynamics – clumsy and inelegant, but it could, sometimes, just get off the ground and fly short distances. The crude Jordan picture was smartened up by Dirac, writing it in terms of electrons so that there was no need to have to bring in vibrating strings any more.

Maxwell's classic equations had shown that electromagnetism has waves travelling at the speed of light. Relativity explained the speed of light. Describing the interaction of matter and electromagnetic radiation clearly had to bring in relativity. In 1926, Jordan and Pauli introduced a toy model of electromagnetism which was relativistic, but contained no electric charges.

Dirac's designer equation for the electron for the first time convincingly linked quantum mechanics and relativity. But Dirac's first attempt at extending his static equation to electrodynamics was clumsy, merely adding conventional electromagnetic terms *à la* Maxwell to his equation. It was like strapping a quantum jet engine to an electromagnetic biplane. It flew, but unpredictably.

HOW TO THROW AWAY INFINITY

In 1929, the powerful duo of Heisenberg and Pauli teamed up to develop a quantum formulation of electromagnetism. Like Dirac and his equation, at first they turned it into a purely mathematical problem, experimenting with a kit of equation parts to see what was needed. Pre-quantum pictures of the electron had always portrayed the particle as a tiny sphere, with electric charge smeared over its surface. In the quantum picture, there was no room for an electron radius, as this would mean having to describe what was happening inside the electron. The electron had to be an infinitesimal point with zero space dimensions but which nevertheless weighed 1/2000 of the mass of a hydrogen atom, carried a negative electric charge and could spin on its axis.

Heisenberg and Pauli found that the zero of the electron's space dimensions arrived in the denominator of their calculations. Anything divided by zero is infinity, and, try as they might, their equations kept throwing up infinities – the interaction of the electron with itself – which jammed up their calculations. It was redolent of attempts in the late nineteenth century to describe the electron as a corpuscle, whose energy density is inversely proportional to its radius. The corpuscle cannot become infinitely small otherwise its energy density becomes infinite. Unlike charges repel, so in this pre-quantum picture the 'pieces' of electric charge inside the electron repel each other more strongly as the electron becomes smaller, and in 1906 the French mathematician

Henri Poincaré even introduced the idea of an internal electron cohesive force to counteract this. Dirac quashed these speculations, saying 'the electron is too simple a thing for the questions of the laws governing its structure to arise'.

The quantum electrodynamics pioneers simply threw away the troublesome infinities and carried on regardless. The puzzle was that the remainder of the equation still worked! More infinities came because of interactions with Dirac's infinite 'sea' of negative-energy electrons. These infinities, too, were discarded but still the equations ticked over nicely. These infinities could be graded, according to how many zeros they had in their denominators. The more denominator zeros there were, the faster the equation blew up and the quicker the infinity had to be got rid of. The quantum electrodynamics aeroplane was a strange vehicle. Major components quickly fell off but the remaining bits still flew!

In 1932, the discovery of the positron made everybody sit up and and try harder. However, Heisenberg was fast losing heart, and spoke of the 'swindles' of having to throw away infinities and pretend that nothing had happened. He was also sceptical that something as complicated as Dirac's infinite 'sea' could ever be accurately formulated mathematically. Nevertheless, he persevered, along with Pauli and Dirac, producing a prototype version of a quantum 'sea' which could produce holes, the positrons.

Victor Weisskopf, a young Austrian student of Pauli, made the first calculations of the interaction of the electron with itself in this new theory, and came up with a whole bunch of infinities. Grading these infinities had now developed into a fine art. But one of Weisskopf's smelt pestilentially. It did not look good for the theory. After Weisskopf published his ominous result, he got a letter from Wendell Furry at the University of California in Berkeley, saying he had done a similar calculation and had found an 'ordinary' infinity. Weisskopf looked again at his calculations and saw where he had gone wrong. In disgust and shame he went to his supervisor, Pauli, and asked if he should give up physics. The arrogant Pauli smiled benignly. 'No', he said, 'everyone makes mistakes. Except me.'

Reassured, Weisskopf returned to the delicate business of grading and discarding infinities, and in a classic 1936 paper wrote down rules for how it should be done. But, as more calculations were made, additional spanners were found in the works. Contributions from low-frequency radiation quanta produced still more infinities, and an artificial frequency 'cutoff' had to be imposed to keep the equations under control. Pauli quit 'in disgust', and, as the clouds of war began to grow, contact between Germany and the rest of the scientific world eventually ceased.

'IF I WERE AN ELECTRON . . .'

In the 1930s, Richard Feynman, a teenage boy in New York's Atlantic Coast suburb of Far Rockaway, preferred collecting mathematical equations rather than postage stamps. Ten years later, this mathematical genius was to transform quantum electrodynamics into one of the most accurate theories ever constructed. Feynman was to quantum electrodynamics what Dirac had been to quantum mechanics. But there the similarity between Dirac and Feynman ended. Dirac was withdrawn and uncommunicative. Feynman was brash and loud. Dirac as a child had been ignored by his father. The young Feynman's intellect had been continually stimulated by his parent. But both these men had an uncanny ability to conjecture how electrons worked. However, while Dirac and the Europeans explored the electron through mathematical formulae, Feynman thought of electrons as his friends. A student friend surmised that Feyman, in thinking about an electron, simply said to himself – 'If I were an electron, what would I do?' Later, Feynman himself supplied the answer – 'The electron does anything it likes. It just goes in any direction at any speed, forward or backward in time, however it likes, and then you add up all the amplitudes and it gives you |the answer|'. Faced with such an unruly particle, Feynman's genius formulated a mathematical framework to handle it.

As well as being a gifted physicist who could strip down any natural problem to its basic parts, see how they were connected and then put them back together, Feynman was a mathematical wizard who could visualize the behaviour of obscure functions in his mind, watching a series expand or seeing an equation generate a curve. Like a jobbing mechanic carrying his own set of spare parts for any eventuality, Feynman was a fanatical collector of mathematical devices and had accumulated an enormous inventory.

Feynman's problem-solving ability was not limited to mathematics. As a boy in the depression, when people would always try to repair something before throwing it away and buying a replacement, Feynman had a reputation for being able to fix gadgets. He was particularly good with radios, then a new technology which few people understood. He had gone to see one client whose radio roared whenever it was switched on. The client, seeing that the 'repairman' was only a young boy, was sceptical about Feynman's abilities. Feynman switched the radio on, heard what happened, and started to walk up and down, thinking about it. 'How do you know about radios? – you're just a boy', taunted the brash client. Feynman quietly interchanged two of the valves in the set and switched it on again. It worked perfectly. Afterwards Feynman got a lot more radio repair work through that man, gaining a reputation as the boy 'who could fix radios by thinking!' One doubts whether Dirac would have been able to fix radios, or, if he had, whether he would have been willing to sell his services. Feynman knew he was different, and bragged about his abilities. But most of the time his bragging could be tolerated.

Feynman was an undergraduate at the Massachusetts Institute of Technology, MIT, which despite its name had a strong tradition in basic science. In his final MIT year, Feynman entered the infamous Putnam mathematics competition for a scholarship at Harvard, where the questions were so hard that the average mark was zero. Feynman walked out of the exam before the allotted time was up and his score was way ahead of the next best candidates. But he had already decided to go to Princeton.

At Princeton, Feynman met the urbane John Wheeler. Just as Ralph Fowler had correctly appraised Dirac and set him on the right path, Wheeler quickly recognized Feynman's ability and pointed him at the most difficult problems from the outset. Feynman had read Dirac's book *The Principles of Quantum Mechanics* and had been excited by its mysterious concluding paragraph: 'It would seem that we have followed as far as possible the path of logical development of the ideas of quantum mechanics as they are at present understood. The difficulties, being of a profound character, can only be removed by some drastic change in the foundations of the theory.'

Feynman resolved that he would be the one who would find whatever Dirac was talking about. But solving Dirac's challenge was to require more thought than fixing rudimentary valve radios. With Wheeler, Feynman began to take apart and re-examine the basic ideas of particle interactions. Light travels fast, but its speed is not infinite. Light from the Sun takes eight minutes to reach the Earth. Electromagnetism needs tiny steps of time to get even from one electron to another, and Feynman built this idea of 'retarded waves' into his emerging picture of electron interactions. Then he noticed something remarkable. If the sign of the time were changed, so that future and past were interchanged, his equations still worked. In doing this, his equations were describing waves which arrived somewhere before they had been emitted! Feynman filed the idea for future use along with all his other mathematical bits and pieces, for the United States had entered the Second World War, and skilled scientists were suddenly in demand.

Feynman was recruited for the theory group at the secret new laboratory at Los Alamos where the atomic bomb was being developed. The main task of the theory group was to predict what would happen in all sorts of unexpected conditions – how quickly uranium gas diffused, how fast neutrons moved, how quickly an atomic bomb would detonate. This meant doing huge calculations, but in those days there were no real computers to help, only mechanical calculating machines. The Head of the Los Alamos Theory Group was Hans Bethe, a gifted physicist who had been one of the first to realize that the Sun is a furnace powered by nuclear reactions. Bethe left Nazi Germany, eventually to work at Cornell University in upstate New York. With Bethe, Feynman was surprised to find someone who could calculate better than he

FIGURE 6.1 Richard Feynman (photo *CERN Courier*). Richard Feynman provided a new understanding of antiparticles.

could. During these war years, Feynman's already formidable mathematical repertoire increased further. With Feynman, Bethe realized he had found an incomparable talent, and offered Feynman a job at Cornell immediately the war was over. Another European emigrant scientist, Eugene Wigner, said that Feynman was 'a second Dirac, only this time human'.

One of the first events at post-war Cornell was the university's bicentennial. The Physics Department organized a three-day seminar on the future of nuclear science, and one of the invited speakers was Paul Dirac. Dirac was still admired, but he was no longer a young man and his ideas had stagnated. At Cornell he repeated his warning that the 1930s formulation of quantum electrodynamics required a radical rethink, but had no suggestions how this should be done. Feynman, delegated to introduce Dirac's talk and lead the subsequent discussion, made a few flippant jokes which did not go down well in such austere company, but reiterated Dirac's call – 'we need an intuitive leap at the mathematical formalism, such as we had in the Dirac electron theory'.

Throughout the history of physics, one technique that always works, although it is not always the most evident when solving a new problem, is what is called the 'principle of least action'. Every motorist knows that it is often quicker to take a longer route using a fast road than a 'short cut' on minor roads. The speed on the fast road makes up for the slow progress on minor roads. But there comes a limit when the detour to get on to the fast road outweighs the extra speed. In planning a cross-country trip, the optimal route can be a complicated series of minor roads and fast roads. So it is with physics, where the 'principle of least action' means finding the optimal solution faced with many alternatives. Nature somehow knows this solution instinctively. The smoothly curved trajectory of a ball in the air is the optimum route that gives the ball the least work to do against gravity. The principle of least action is not the best way to calculate the trajectory of a ball, but it is an elegant way of understanding it. The principle of least action is the Rolls-Royce of mathematical physics.

In the late 1930s, Dirac had turned to this approach in his search for a new attack on quantum electrodynamics. Dirac's reformulation of quantum mechanics in 1925 had relied on close parallels between quantum mechanics and the detailed formalism of classical mechanics – his Poisson brackets. Poisson brackets had originally been tailor-made for the principle of least action. More than ten years after his Poisson bracket analogy, Dirac had searched for a new mathematical analogue to guide quantum electrodynamics. But the sudden inspiration he had on that Sunday walk in 1925 had not come. Instead, in 1933 he made a vague suggestion as to how quantum electrodynamics might parallel the time-honoured principle of least action. But it remained as a vague suggestion and he had not followed it up.

During Cornell's 1946 bicentenary celebrations, Feynman looked out of the room where the nuclear physics symposium was being held and saw Dirac lying on the grass. Feynman had read Dirac's old suggestion and had thought deeply about it, and had realized what Dirac was driving at in his 1933 paper. For Feynman, this was the way ahead. On the Cornell lawn he approached the older physicist and asked whether he had a direct mathematical bridge between the least-action recipe and quantum electrodynamics. Dirac looked surprised, and walked away.

To focus his own thinking, Feynman had completely reformulated Dirac's picture of quantum mechanics using this least-action principle. The idea was to find all the possible things a particle could do in any situation, add them up and calculate the optimal one. Initially, there was no relativity in it, so it gave nothing new. But Feynman was proud of it. 'There is a pleasure recognizing old things from a new point of view', he said.

THE LAMB SHIFT

To focus thinking on the problems of quantum electrodynamics, in 1947 a small brainstorming workshop was organized at Shelter Island, on New York's Long Island. All the big names from the Los Alamos days were there for a programme of talks organized by Victor Weisskopf. In the middle of the Shelter Island meeting came a scientific bombshell. At New York's Columbia University, a young researcher named Willis Lamb had shone microwaves on hydrogen and found that two separate energy levels, which according to the Dirac equation should have been exactly equal, were instead separated by a tiny energy gap – what came to be known as the 'Lamb shift'. Instead of focusing on the problems of recurring infinities and figuring out how to handle them, the Shelter Island meeting was stunned by this new development. Here was an unsuspected chink in the all-powerful Dirac equation. Unfazed, Hans Bethe, on the train home from the meeting, did a traditional messy old-style quantum electrodynamics calculation on the back of an envelope, throwing away a few infinities and arbitrarily cutting off some wavelengths.

He got the right answer for the Lamb shift, but Bethe knew this was not the way for quantum electrodynamics to go. Back at Cornell, Bethe threw the problem at Freeman Dyson, a brilliant young student from

FIGURE 6.2 Mathematical perfection and sophistication – Julian Schwinger (Harvard University, courtesy AIP Emilio Segré Visual Archives).

England who had written a science-fiction novel at the age of eight and subsequently made a reputation for himself as a mathematician. At the other end of New York State, at Columbia University, where Willis Lamb had done his experiments, a young physicist called Julian Schwinger already had a mathematical solution.

The same age as Feynman, Schwinger, while still at high school, had gone into the library at the City College of New York and read papers by Dirac. During the war, Schwinger worked on radar, a less glamorous, but no less important, aspect of the US war effort. Suave and urbane, Schwinger aimed for perfection and sophistication in his work. However these qualities frequently made his ideas difficult to follow. Some thought he did this on purpose. Schwinger's new approach deftly side-stepped the infinities which had plagued Bethe, Dirac, Heisenberg, Pauli and Weisskopf, and it gave the right answer for the Lamb shift.

In Cornell, Feynman had been continuing with his least action ideas, trying to incorporate the ideas of relativity. Feynman remembered the idea he had tried out at Princeton with Wheeler, with the waves arriving before they were emitted by the oscillators, waves travelling backwards in time. Feynman found that if these possibilities were included in his least-action picture, suddenly everything came right. A new picture of the relativistic electron was born. The positron, said Feynman, is an electron travelling backwards in time!

Feynman struggled with the novelty of the idea, which meant abandoning the idea that the future can have no impact on the past, at least on the microscopic scale. But it was relativity at work. Einstein had showed that there was no absolute zero of time, all times were relative. So why not future times too? 'It may prove useful in physics to consider events in all of time at once and to imagine at each instant we are only aware of those that lie behind us', said Feynman. He even had a rough-and-ready analogy. A tourist on foot will simply walk up a hill. However, a motorist has to follow another route. Planning a route through a mountainous region from a map, a motorist following the path of one road suddenly sees several roads running in parallel. To climb the mountain, the road has to switch back on itself and take a zig-zag route to the top. In planning his optimal route, the motorist cannot go in a straight line – all these switchbacks have to be taken into account. Before setting out, a quantum has to worry about what it will do in the future.

The ideas, as applied to electrons and positrons, could be written as compelling little pictures – 'Feynman diagrams' – subway maps of sub-atomic physics. The electrons and positrons are lines, but have arrows, showing in which direction time runs. Photons, the quanta of radiation, look the same whether they are running forwards or backwards in time, and are shown as wiggly lines. The three dimensions of space and the fourth dimension of time cannot be accommodated on a two-dimensional sheet of paper. Feynman diagrams (Figure 6.3) represent a simplified world of one space dimension and one time dimension. But the space and time dimensions can be interchanged – on this relativistic map, north–south travel is just as easy as west–east. By turning the map, physics processes that pre-1947 were viewed as totally different

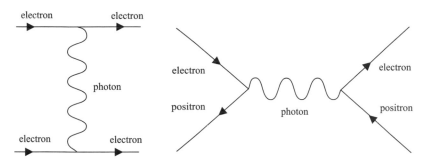

FIGURE 6.3 Feynman diagrams represent a simplified world of one space dimension (up and down) and one time dimension (left and right). On the left is the scattering of an electron by another via the exchange of a quantum of radiation, a photon. But the space and time dimensions can be interchanged. Turning the diagram shows a related reaction, an electron and a positron annihilating into a puff of radiation, later rematerializing as another electron–positron pair. The positron appears as an electron with its time arrow reversed.

now became intimately related. Thus the scattering of an electron is the cousin reaction of an electron and a positron annihilating into a puff of radiation, later to rematerialize as another electron–positron pair. For electrons, time travel had arrived.

Feynman's time arrows provided a compelling new understanding of the positron, avoiding Dirac's painful sea of negative energies. But there was no side-stepping the complexities of the vacuum. Instead of having to be viewed as all possible interactions with the infinite negative energy sea, the vacuum now became filled with little bits of Feynman diagrams, electron–positron pairs popping on and off, or stray bits of photon looping through space-time before disappearing back into a parent electron (Figure 6.4).

Feynman's graphic approach and Schwinger's urbane algebra provided the same explanation for the Lamb shift. But, while Schwinger's formulation was difficult to follow, Feynman's little diagrams let physicists keep track of what was happening. Schwinger later said that Feynman's diagrams 'like today's silicon chip, brought calculations to the masses'. (Many people have interpreted this statement as an

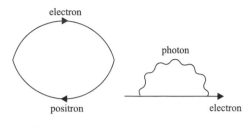

electron

positron

photon

electron

FIGURE 6.4 Vacuum pictures. The physics vacuum is filled with little bits of Feynman diagrams, electron–positron pairs popping on and off, or stray bits of photon looping through space-time before disappearing back into a parent electron.

appreciation of Feynman's work, however, the sophisticated Schwinger was privately derisory of this 'easy' approach.)

A new word appeared in the physics vocabulary – 'renormalization' – to describe the new way of dealing with unwanted infinities. These unwanted infinities always crop up in calculating basic quantities, like the electric charge or the mass of the electron. The cunning idea is to *choose* the value of the basic charge or mass, which has to be fed into the equations anyway, in such a way that the total charge or mass which comes out is equal to the experimentally observed value, which is clearly not infinite. The total charge or mass fed into the equations includes an infinite part, this infinity being equal and opposite to the infinity produced by the calculations. This recipe, absurd at face value, nevertheless works, because the unwanted infinities always occur when dealing with quantities, like the electron's mass and electric charge, which cannot be predicted from scratch and have to be fed into the equation by hand. Why not feed in infinities?

Whether one preferred the everyday Feynman model or the de luxe Schwinger edition, quantum electrodynamics had finally arrived and everyone clambered aboard. Calculations were no longer blocked by infinities and Dirac's artificial idea of an infinite sea of negative energy electrons, so painfully invented and so difficult to work with, could be discarded. Future generations of physicists would learn quantum electrodynamics straight from Feynman diagrams and would wonder what all the fuss about a 'sea' had been about. These new ideas had also been developed independently in Japan by Sin-ichiro Tomonaga. In

1965, Feynman, Schwinger and Tomonaga shared the Nobel Prize for Physics.

Throughout his life, Feynman was colourful and controversial. Many thought him obnoxious, but he had a noble side too, shown none more so than in his first marriage, to Arline Greenbaum, who he married secretly in 1942, knowing she was mortally ill with lymphatic tuberculosis. At the wedding ceremony, he did not kiss his bride for fear of catching the disease himself. He cared for her lovingly until her death in 1945. One month after her death, he noticed a woman's dress in a shop window which struck him would have looked good on Arline. He broke down in the street and cried.

In 1981, Feynman underwent surgery for abdominal cancer. In August 1981, Feynman and I attended an international physics conference in Lisbon, Portugal. One morning, we had both decided to skip one particularly boring talk and were sitting in the cafeteria. I introduced myself to him and asked about his health. Without even answering my question, he was rude and insulting. Had the setting been different, I would have insulted him back or shouted at him. But this was the great Richard Feynman. As I turned round and walked away, I heard him laughing. I did not mention this episode when I wrote his obituary, 'Physics in the Fast Lane', in 1988.

7   The quark and the antiquark

Light, produced when electric matter vibrates but itself carrying no electric charge, is a very special kind of matter. Antilight, produced when electric charges of antimatter vibrate, is the same as light. The antiworld is lit by the same light as our own. But twentieth-century physics has discovered that matter has more labels than electric charge. In antimatter, these additional labels are reversed: many electrically neutral particles can have antiparticles.

After Rutherford discovered the nucleus and showed that it contained protons, he soon realized that protons were not all the nucleus contained. Protons are two thousand times heavier than electrons, so the contribution of electrons to the mass of the atom is almost negligible. But simply adding up the number of protons gave the wrong result for the mass of the nucleus. Only about half the mass of the atom could be accounted for by its electrically charged nuclear protons. The missing mass, said Rutherford, is due to more nuclear particles, about as heavy as protons but electrically neutral. These he called neutrons.

Bombarding light atoms like beryllium with alpha particles, in 1930 the German physicists Walther Bothe and Herbert Becker produced something that could pass through 10 centimetres of lead. They first thought it was some kind of radioactivity, but in 1932 Irène Curie and her husband Frederick Joliot showed that whatever it was could knock protons out of hydrogen. After Pierre and Marie Curie's epic investigations of radioactivity in 1898, this was the second-generation Curie husband-and-wife team to make a major scientific breakthrough. But they did not fully investigate its implications. Pushed by Rutherford, the Cambridge team swooped in to finish off the job. James Chadwick, another Rutherford super-student who went on to win a Nobel prize, looked at the Joliot–Curie protons and found that they behaved as though they had been liberated by particles of about the same mass as the proton. Here were Rutherford's neutrons.

Most nuclei contain about as many neutrons as protons, the neutrons acting as a nuclear counterweight to balance the immense electrical repulsion when positively charged protons are crammed together – about ten billion times stronger than the attraction which binds a proton to an atomic electron. The nuclear superglue between protons and neutrons is usually strong enough to resist the forces trying to push the protons apart. But not always.

On 20 December 1938, a Christmas party at the Kaiser Wilhelm Institute, Berlin's major scientific laboratory, should have marked the end of research work for that year. However, two radiochemists, Otto Hahn and Fritz Strassmann, had been busy showering uranium with neutrons. They knew that new radioactive products were produced, but had not been able to identify what these product nuclei were. Too impatient to accept a Christmas break, the scientists continued their experiments in the semi-deserted laboratory. They expected that a uranium nucleus would swallow an incoming neutron and transform into a kindred heavy nucleus, losing a few fragments as it did so. Instead, they were astonished to find that they had catalysed a new kind of nuclear process. A uranium nucleus was unable to digest the extra neutron and split into two roughly equal halves, emitting a few more neutrons as it did so. The new process was called nuclear fission and the emitted neutrons could catalyse more fissions – a chain reaction. Within seven years this new neutron chemistry would change the course of world history. On 16 July 1945, the first atomic bomb was tested near Almagordo in the New Mexico Desert. Within a few weeks, bombs were dropped on Hiroshima and Nagasaki and the Second World War ceased.

Left to itself, the innocent neutron is a passive sentinel of nuclear stability. But in the depths of space, neutrons play a different role. Giant stars act as cosmic vacuum cleaners, their fierce gravity remorselessly sucking in stray wisps of interstellar gas. Such a star gradually gets larger and heavier, eventually reaching a stage where the crush of its gravity overcomes the springy forces in its constituent atoms. Under this gravitational compression, atoms collapse. The orbital electrons in the star's atoms are pushed into the nucleus, each electron's negative electric charge cancelling the positive electric charge of a nuclear proton, giving a neutron. Without any atomic padding, the resulting 'neutron star' is only about ten kilometres across, but this dense nuclear matter is more than a million million times heavier than conventional atomic matter. Each cubic centimetre of neutron star weighs about a trillion tons. In the depths of outer space, these compact stars, composed entirely of neutral nuclear particles, have lost the electrical imbalance of ordinary atomic matter. If a neutron and an antineutron were one and the same particle, like a particle of light, neutron stars would be both matter and antimatter, equally at home in our world or an antiworld. But it turns out that a neutron is not its own antiparticle. The neutron has an inner structure which is recognizably different in the antimatter mirror. A neutron star is not free to enter the antiworld.

ATOM SMASHERS

Those who helped develop neutron physics during the Second World War – including Nobel prizewinners Hans Bethe, Enrico Fermi, Richard Feynman, Ernest Lawrence – had to resolve a terrible personal dilemma between their patriotic duty and their conscience at perfecting such weapons of mass destruction. With the war over and their patriotic duty done, they clamoured to quit the weapons programme and return to their universities to carry on with pure research. While their wartime physics had changed the world, the world had also changed physics. Developing the bomb had required an effort of major industrial proportions, and physicists had acquired new managerial skills. These scientists also emerged as the heroes of the Second World War, magicians who could conjure devastating explosions from tiny fragments of matter and bring conflicts to an end. As a recompense, and to retain the nuclear advantage in the ensuing Cold War, the physicists could have whatever they asked.

In the early 1930s, the American physicist Ernest Lawrence, working at the University of California campus at Berkeley, on the hills above the San Francisco Bay, had invented the cyclotron, a machine for accelerating charged particles to high energy. Injected at the centre of Lawrence's circular device, nuclear particles spiralled round faster and faster in a special arrangement of electric and magnetic fields, an electromagnetic slingshot. Lawrence's new subatomic accelerator provided a powerful new hammer to break up atomic nuclei. The 'atom smasher' had arrived, and Lawrence earned the 1939 Nobel prize. The Berkeley laboratory and Lawrence went on to play vital roles in the wartime effort. The Berkeley staff grew to almost 1,200, including 65 security guards. Post-war, the laboratory inherited a lot of money and resources from the wartime effort. Other universities rushed to build the new cyclotron status symbols. On the other side of the continent, a major new laboratory was set up on an old army base on Long Island, near New York. Camp Upton, originally established in 1917 as a staging post for US troops *en route* to First World War battlefields in Europe, became the Brookhaven National Laboratory.

While this new post-war scientific effort was being marshalled in the US, in Europe physics was still geared to its modest pre-war scale. With no major new construction projects, at least initially, nimble researchers carried on where they had left off. As if trying to make up for lost time, dramatic discoveries were made by eager physicists who climbed mountains to expose photographic plates to the increased levels of cosmic rays at high altitudes. Before the war, Europe had been the acknowledged centre of physics, and a spell at a major European university was almost mandatory for aspiring US researchers. This pattern was interrupted when European scientists went to North American to collaborate in the atomic bomb project, supported by the immense resources of the US. Worried that their newly acquired status might slip from their grasp, American physicists watched enviously as European physicists discovered a series of new particles – the mu-meson or muon, the pi-meson or pion, so-called 'V particles', the tau meson . . . Had the physics pendulum swung back to Europe? Would the new US cyclotrons be white elephants?

In April 1948, the US Atomic Energy Commission authorized the construction of two giant new atom smashers, one at Brookhaven and the other at Berkeley. In 1952, Brookhaven's 'Cosmotron' came on line, and was soon mass-producing the new particles first seen in European cosmic ray experiments. The Europeans had to stand aside as the US physics express train came through. The Berkeley machine, slightly higher in energy, had a specific aim – to find the antiproton, the anti-particle counterpart of the proton, which had been boldly predicted by Dirac in 1932. However, not everyone was optimistic about the antiproton, and wagers were made.

THE ANTIPROTON

Cosmic ray specialists had hoped that the antiproton, like the positron, would also fall into their net, but the essence of nuclear antimatter proved much more elusive. In Dirac's language, a quantum of radiation had to have enough energy to dig a particle out of the 'sea' of negative energy states. By 1947, this argument could be translated into more compelling language – a quantum of radiation had to transform into a particle–antiparticle pair. To accomplish this, the quantum of radiation had to supply twice the mass of the particle or antiparticle alone. The cosmic ray quanta seen by Blackett and Occhialini only had to make two electron masses. To make a proton–antiproton pair, the radiation had to be two thousand times more powerful. Finding such energies in cosmic rays was difficult. In 1954 and 1955 came hesitant cosmic ray reports of the elusive antiproton, but the results were inconclusive. When the new Berkeley machine, the Bevatron, started up in 1954, it was the world's most powerful atom smasher. The scene for the antiproton had been set.

Anderson's pioneer experiments had shown how difficult it was to distinguish between positrons moving one way and electrons moving the other. At Berkeley, Owen Chamberlain, Emilio Segrè, Clyde Wiegand and Tom Ypsilantis began to prepare a trap for antiprotons. Segrè, born in Rome in 1905, had worked in the 1930s with Enrico Fermi at the University of Rome. Fermi had gone to Stockholm in 1938 to collect the Nobel Prize for Physics and never returned to Fascist Italy. Instead he made his home in the USA. Italy's loss became the USA's gain. Fermi went on to lead the team which built the world's first nuclear reactor at the University of Chicago, and subsequently moved to the atomic bomb project at Los Alamos. Segrè followed Fermi to the US, and became a group leader at Los Alamos. Wiegand had previously worked with Segrè at Berkeley and Los Alamos. Chamberlain, a Californian, worked with Segrè at Los Alamos. After a post-war spell at Chicago with Fermi, Chamberlain moved to Berkeley. Ypsilantis, a young postdoctoral researcher, had recently joined the Berkeley faculty.

The four experimenters patiently waited for the Bevatron to reach its design energy and peer over the antiproton wall. They knew that about one proton in every million produced by the new Bevatron could go on to produce an antiproton, the other protons producing sprays of other particles. To separate the precious antiproton wheat from the chaff, they used an ingenious system of magnetic lenses. In the same way that a prism splits a ray of white light into its component colours, passing a mixed particle beam through a magnetic field separates different energies. Just as Anderson had used the curvature of tracks in a magnetic field to differentiate between fast and slow positrons, so the Bevatron experiment used a magnet to sweep away positively charged particles and filter off a hopefully antiproton-rich beam of negatively charged particles. In fact they used two magnetic prisms in series, one after the other, to further refine the beam.

To identify antiprotons, the experiment measured the time it took for the particles to travel between the two magnetic lenses, 12 metres apart. Most of the subnuclear particles produced by the Bevatron would travel almost at the speed of light and would take 40 nanoseconds (a nanosecond is a thousandth of a microsecond) to cross the 12-metre gap. Antiprotons, being much heavier, would travel more slowly and take 51 nanoseconds. The ability to detect antiprotons relied on the ability of 1950s electronic circuits to detect time differences of 11 nanoseconds. But even picking up pairs of counts separated by the required 11 nanoseconds would not be enough. It could happen that two unrelated particles in close succession could have just that time interval between them, and could 'fool' the electronic circuits into thinking that an antiproton had passed.

To close this loophole, the team proposed also measuring the velocity of particles by a separate method. Although nothing can travel faster than light in a vacuum, a high-energy particle can drill its way through a transparent solid such as glass faster than the natural speed of light in glass. When this happens, an optical shock wave is set up, called Cherenkov light, after the Russian physicist Pavel Cherenkov. The direction of this light depends on the velocity of the high-energy particle. In the experiment's ingenious velocity filter, a cylindrical mirror ensured that only light corresponding to antiprotons was reflected on to a focus, where it was picked up by a photomultiplier.

In 1955, the Bevatron energy reached the antiproton threshold. The electronics was set to 'go off' each time a delay of 51 nanoseconds was detected, and at the same time the Cherenkov photomultipliers recorded light. Eagerly the four experimenters watched their oscilloscopes. Nothing happened. No tell-tale antiproton blips were seen. Sensing that they may have made an error and spoiled the delicate settings, they reversed the field in their magnetic prisms, which would make their apparatus sensitive to the copious positively charged protons. Seeing no protons, they checked their calculations and found a wrong setting in their magnetic prisms. After correcting it, they started with a proton check, and only after seeing lots of proton counts did they reverse the magnetic field and select antiprotons instead. Soon the first antiproton candidate counts were logged. Those who had bet against antiprotons had to pay up. The experiment continued to log antiprotons for three months and was able to show that the proton and antiproton had the same mass to within some 5 per cent.

The discovery was soon confirmed at the Bevatron in an experiment which revealed the antiproton for all to see. Instead of a sophisticated electronic detector, the second experiment substituted a slab of photographic emulsion of the sort that European teams had used to record

FIGURE 7.1 Members of the team that discovered the antiproton – left to right Tom Elioff, Bob Backenstoss, Rudy Larsen, Clyde Wiegand, Tom Ypsilantis. (Photo Tom Ypsilantis)

their earlier cosmic ray discoveries. In the team was Edoardo Amaldi, an Italian physicist who had once caught a fleeting glimpse of a possible antiproton in a cosmic ray exposure, but could not be sure. The dramatic emulsion 'star' of antiproton production at the Bevatron quickly became a subnuclear physics collector's item. In 1959, Segrè and Chamberlain shared the Nobel Prize for Physics, one year after Pavel Cherenkov, whose discovery of the radiation which bears his name had been so important for their experiment.

No sooner had the antiproton been discovered than it became a stepping-stone to the next antiparticle. In 1957, Bill Cork, Glen Lambertson, Oreste Piccione and Bill Wentzel set out to look for the antiproton's antinuclear partner, the antineutron. This would be more difficult. Carrying no electric charge, the antineutron cannot be detected directly, only indirectly, through its interactions. The Berkeley team tamed antiprotons and shone them into a bath of liquid scintillating material. Subsequent counters registered whether a charged particle passed, and finally the beam met the glass block of a Cherenkov counter. The experimenters saw seventy-four examples where an incoming antiproton apparently lost its charge in the initial scintillator target, with no charged particle passing through the next sensitive counters, but with the Cherenkov detector picking up the results of an interaction. In these interactions, the antiproton had surrendered its charge in the first target, transforming into an antineutron, which passed unseen through the charged particle detectors until it crashed into a nucleus of the Cherenkov glass, annihilating with a subnuclear particle and producing characteristic flashes of subnuclear debris.

With the antiproton and the antineutron in the bag, the next step was to manufacture an antinucleus. The first piece of nuclear antimatter was synthesized by the ambitious Italian physicist Antonino Zichichi in 1965. Hydrogen, the simplest atom of all, is made up of a lone electron orbiting round a lone proton. Its nucleus is just a single particle. However, about one naturally occurring hydrogen atom in ten thousand is different – its nucleus still contains only one proton (otherwise it would no longer be hydrogen), but in addition a stray neutron has locked on to the proton. With an atomic mass of two units, this 'heavy hydrogen' is called deuterium, and the proton–neutron nuclear pair, the simplest compound nucleus, is termed a deuteron. Water made with deuterium, D_2O, instead of the usual H_2O, is 10 per cent heavier than ordinary water – 'heavy water'. Working at the European CERN laboratory in Geneva and using a highly refined beam of negatively charged particles to boost the antiproton supply, Zichichi found antideuterons, each composed of an antiproton and an antineutron. Antinuclei can stick together in the same way as nuclei.

PARTICLES AND ANTIPARTICLES APLENTY
In the late 1950s, the Brookhaven Cosmotron and the Berkeley Bevatron were joined by even higher energy accelerators – a bigger machine at Brookhaven and another at Argonne, near Chicago. These huge new

atom smashers cost a lot of money, more than any one European nation could afford, so, to keep pace with the Americans, the nations of Western Europe had clubbed together to form CERN and built a big machine of their own in Geneva, Switzerland. This new generation of proton accelerators reaped an unexpected harvest of new particles. Wherever the experimenters looked, it seemed there was a new particle to be found. The initial objective had been to explain the structure of the nucleus, but this was soon overlooked in the gold rush to find new particles. With universities expanding rapidly, there was no better ticket to a professorial chair than to stake a claim on a new particle.

All these new particles were highly unstable, living for a tiny fraction of a second before decaying into lighter ones. But there were simply too many particles. It was like the bull market in new chemical elements in the mid-nineteenth century once the necessary chemistry had been perfected. Then Dmitri Ivanovich Mendeleyev had brought order by showing that arranging the elements into a grid structure revealed eerie similarities between elements which at first sight are very different – fluorine, a gas, is a close relative of iodine, a solid. The reasons underlying this pattern of elements only became clear when the electronic structure of the atom was revealed in the early twentieth century. The regularities in Mendeleyev's table reflected the quantum seating plan for atomic electrons. What lay behind the rash of 1950s particles? What creates so many particles and what makes them subsequently decay?

Murray Gell-Mann played the same role for particle physics that Mendeleyev had for elemental chemistry. Born in New York City, Gell-Mann was first noticed as an intellectually precocious teenager. A genuine polymath, he has a keen ear for phonetics and language. Even his own surname, with its equal stress on the two syllables, reflects this care. Only slowly did he gravitate towards a physics career, eventually joining Enrico Fermi's school in Chicago in the early 1950s. There, Gell-Mann decided to investigate the epidemic of unstable particles and search for some underlying order. Faced with a baffling complexity, he looked for simple underlying regularities.

Animals come in all shapes and sizes, but they can be usefully classified by how many legs they have. Fish and snakes have no legs, humans and apes have two, most mammals four and insects six. Looking at the way animals reproduce, this 'legginess' is important – in order for a union to be productive, an animal has to mate with a partner of equal legginess. For reproduction, one can say that legginess has to be 'conserved'. However, on the completely different time-scale of evolution, legginess plays no role – the ancestors of all living animals came from the sea and had no legs at all. So it could be with particles, reasoned Gell-Mann, who looked for a particle equivalent of legginess. Electric charge is always conserved, but there was a somewhat analogous quantity, which physicists called 'hypercharge'. Each subnuclear particle had both an electric charge and a hypercharge label. Electric charge is always conserved, but hypercharge is only conserved on the brief time-scale when particles are formed, not on the much longer time-scale when they decay.

Gell-Mann's taste for words did not like the name hypercharge, and he invented a related label – 'strangeness'. The unstable particles being manufactured by the new proton accelerators were indeed strange, so what not call them so? But more conventional physicists did not like the idea of strangeness being a measurable quantity. Used to calling basic particles -ons and naming new ones after the letters of the Greek alphabet, they thought Gell-Mann's proposal too whimsical and lacking in seriousness. His proposed title 'Strange Particles' had to be edited to 'New Unstable Particles'.

In 1954, Gell-Mann moved from Chicago to Caltech. By this time more than thirty unstable subnuclear particles were on the books, most carrying Gell-Mann's 'strangeness' label. Applying ideas of mathematical symmetry to produce two-dimensional patterns, Gell-Mann saw that when particles were labelled by their electric charge and strangeness, they fell naturally into families – geometrical patterns containing eight, or sometimes ten, members. These patterns were the subnuclear equivalent of the up and down clues in Mendeleyev's crossword of elements. In London, Yuval Ne'eman, doing physics research in parallel with his official duties as Israeli military attaché, had the same idea and found the same patterns.

But what lay behind this intriguing new seating plan? Mathematically, the eight- and ten-fold families followed from the different ways of arranging three basic components. This idea was put forward simultaneously by Gell-Mann and by George Zweig, another Caltech researcher. Ne'eman too realized the significance of the underlying triplet. Linguist Gell-Mann painstakingly pronounces every proper name authentically, but, for his own science, he opted for a nonsense word for the triplet. In his book *The Quark and the Jaguar* (1994), Gell-Mann explains that when he first had the idea of three basic constituents, he privately used the sound 'kwork'. Gell-Mann's phonetic sensitivity was continually stimulated by the experimental language in James Joyce's book *Finnegan's Wake* (1939). Dipping into the book, Gell-Mann stumbled across the phrase 'Three quarks for Muster Mark'. With the threefold connection, the word 'quark' seemed appropriate for a threefold mathematical structure. But pronunciation was important for Gell-Mann. Should 'quark' rhyme with 'mark', as the phrase suggested? Finnegan's Wake is the dream of an innkeeper called Humphrey
Chimpden Earwicker, and calls for drinks at the bar are an underlying theme in the book. Gell-Mann reasoned that quark was meant to imply 'quarts' – 'three quarts for Muster Mark', so that quark rhymed instead with 'kwork', his original phonetic device. However it was pronounced, for other physicists the name quark was even more unacceptable than strangeness. Such casual terminology would demean the solemnity of their trade, they said, but to Gell-Mann's delight the name 'quark' stuck. However, these days most people mispronounce the word. For Gell-Mann, it should rhyme with 'cork', rather than 'mark'.

One of Gell-Mann's three quarks carried the charge-like quantity, strangeness, that he had introduced ten years before. It was the 'strange quark'. The other two quarks, found in everyday protons and neutrons, were very much a pair, like Tweedledum and Tweedledee. Physicists named them 'up' and 'down'.

FIGURE 7.2 Mr Quark–
Murray Gell-Mann in a London
pub with the typescript of his
book *The Quark and the Jaguar*
(photo Maurice Jacob).

With quarks on the table, their antimatter counterparts, antiquarks, were ripe for picking. Heavy subnuclear particles, like the proton and neutron, behave as though they contain three quarks. The proton has two up quarks and one down one, the neutron one up quark and two down ones. The antineutron has one up antiquark and two down antiquarks, very different to the neutron. Lighter particles, such as the pion, which have no role in everyday nuclear affairs, look as though they are composed of a quark and an antiquark. In the quark world, antiparticles play a very important role. Quarks and antiquarks link together to make a rich array of exotic particles; however, none of these particles are stable, and they are therefore unknown in everyday matter.

The idea of quarks was compelling, but at first physicists were unwilling to accept that the proton and neutron actually contained smaller particles. The quark 'structure', they said, was simply a mathematical trick. However, experiments in the late 1960s and early 1970s actually saw evidence for quarks. It was a rerun of Rutherford's experiment sixty years before, which had discovered the atomic nucleus. Just as Rutherford had seen alpha-particle projectiles recoiling from something hidden deep inside atoms, the new experiments saw their infinitesimally small electrons recoiling from something hidden deep inside protons. Compared to the proton, quarks are as small as the nucleus is to the atom! However, unlike the atom, where most of the space between the orbital electrons and the compact central nucleus is empty, a subnuclear particle like the proton is full of quark debris – quark–antiquark pairs continually popping on and off to spice up the vacuum. Deep inside the proton, antiparticles are always present.

8 Broken mirrors

> 'I'm sure I'll take *you* with pleasure!' the Queen said. 'Twopence a week,
> and jam every other day.'
> Alice couldn't help laughing, as she said 'I don't want you to hire *me* – and I
> don't care for jam.'
> 'It's very good jam,' said the Queen.
> 'Well, I don't want any *today*, at any rate.'
> 'You couldn't have it if you did want it,' the Queen said. 'The rule is, jam
> tomorrow and jam yesterday – but never jam today.'
> 'It *must* come sometimes to jam today,' Alice objected.
> 'No it can't,' said the Queen. 'It's jam every *other* day: today isn't any *other*
> day, you know.'
> 'I don't understand you,' said Alice. 'It's dreadfully confusing!'
> 'That's the effect of living backwards,' the Queen said kindly; 'it always
> makes one a little giddy at first –'
> 'Living backwards!' Alice repeated in great astonishment. 'I never heard of
> such a thing.'
> ' – but there's one great advantage in it, that one's memory works both
> ways.'
> 'I'm sure mine only works one way,' Alice remarked. 'I can't remember
> things before they happen.'
> 'It's a poor sort of memory that only works backwards,' the Queen
> remarked.

(From *Through the Looking-Glass*, by Lewis Carroll)

When Alice was enticed into the mirror world, her experience, limited
to the familiar world on her side of the fireplace, meant she was ill-
prepared for the idiosyncrasies on the far side of the mirror. Because
Feynman had shown that antimatter 'works backwards', as the White
Queen put it, it is natural to think of antimatter, too, as some kind of
mirror world, but the mirror of antimatter is a very special one, and its
surprises would startle even an experienced Alice.

In an ordinary mirror, left appears right and right appears left. A
right-handed screw reflects as a left-handed screw. 'Handedness', like
gravity, is deeply ingrained into human consciousness. Right is traditionally associated with good or being well (are you all right?) while left
('sinister' in Latin) has connotations of bad. A skilled person is 'adroit',

from the French 'droit', meaning 'right' or 'correct', or 'dextrous' from the Greek 'dexios', on the right. A clumsy person is 'gauche', from the French word for left. The Latin adjective 'sinister' meant unlucky and inauspicious as well as left-handed. In politics, the eighteenth-century tradition of the assembly of French nobility to sit with traditionalists on the right of the chamber while their opponents sat on the left is enshrined in the connotations of right- and left-wing.

Living things, having to contend with the pull of gravitation, are very asymmetric vertically. Roots and feet look very different to flowers and heads. Animals also show a marked front–back asymmetry, as they prefer to look in the direction in which they move. However, with no everyday effect distinguishing between left and right, most living things have at least a superficial left–right symmetry. But a second look shows that this is not the case. Charles Lutwidge Dodgson, also known as Lewis Carroll, the author of *Alice's Adventures in Wonderland* and *Through the Looking-Glass*, was more asymmetric than most people, with one shoulder and one eye slightly higher than the other. Inside the body, less visible, the organs are not arranged symmetrically, and, as if to underline this subtle lack of balance, people choose asymmetric hairstyles, while the details of clothing, such as pockets and fastenings, can have some left–right bias. A landscape can be printed the wrong way round without most people knowing, but by examining details of clothing we can usually tell if a portrait has been inadvertently printed back-to-front, even if we do not know the person pictured.

NATURAL HANDEDNESS

The human body is superficially left–right symmetric, but some 90 per cent of people prefer to use one hand rather than the other. Distinguishing left and right can appear to be merely a matter of everyday convention, like pockets and buttons. Objects are assembled using right-handed screws, but would be equally satisfactory if assembled using left-handed screws. A screwdriver works equally well when turned in either direction. If a 'handed' object exists or process happens, then its mirror-image should also exist or be able to happen. Right can be

FIGURE 8.1 Dichloromethane. The atoms of dichloromethane can be arranged in different ways. Seen in a mirror in front of the molecules, the example on the left reflects as itself, that on the right looks upside down.

defined as 'the hand that most people prefer to use', but close study reveals that the idea of handedness is much more deeply implanted in Nature.

The science of stereochemistry looks at the arrangement of the constituent atoms of complex molecules. If the same atoms are stuck together in different ways, the resulting molecules have the same composition but different structures, these structures not being mirror-images of each other. These different atomic arrangements are called 'stereoisomers'. This is particularly the case in complex organic molecules. Dichloromethane, $C_2H_2Cl_2$, is a simple example (Figure 8.1).

Such asymmetry can result in two mirror-image molecules having different properties. Dextrose (glucose) has a right-handed structure. Enzymes can only digest dextrose. Some natural sweeteners use left-handed molecules which are not digested and so do not produce calories. It was realized too late that the drug thalidomide has radically different properties, depending on the left–right orientation of the atoms in its molecules. At the fundamental level, most amino-acids, the building blocks of living material, have a left-handed structure. Another signal that living material cares about direction came with the famous 1953 discovery by James Watson and Francis Crick of the 'double helix' of DNA, the hereditary material of living organisms. The molecules of DNA are arranged in two strands in a ladder-like structure, which is then twisted into a double spiral.

Until the mid-1950s, physicists thought that mirror symmetry worked for subatomic processes. If asked the question, a physicist would have said that Nature can use right-handed or left-handed subatomic screws equally well. But few people had asked the question.

The electron is a spinning particle, and the Dirac equation showed that this spin can point in one of two directions, upwards (spinning clockwise, or right-handed) or downwards (anticlockwise, or left-handed). A clockwise-spinning electron, viewed in a mirror, appears as an anticlockwise electron. Both exist in Nature, and there was no reason to suppose that an electron process and its mirror-image would behave differently. A left-handed electron 'clock', despite looking different to a right-handed clock, would still tick at the same rate.

In quantum physics, the idea of left–right symmetry had been transformed into a law that conserved a charge-like quantity called 'parity', which cannot be accumulated like electric charge. Like an up–down switch, parity can have only two values – 'odd' or 'even'. Physicists assumed that the parity of what went into a reaction, even or odd, was the same as the parity of what came out. Introducing parity sidestepped having to imagine difficult manœuvres with mirrors and provided a useful check on calculations.

In the early 1950s, at the same time as Watson and Crick were telling an astonished world about the double helix of DNA, physicists' attention was focused on two new unstable, electrically neutral particles, which they had named tau and theta. Using physics vocabulary, the tau and the theta were 'strange' – they carried Gell-Mann's additional charge. They decayed in very different ways, and had different parities, even and odd. With everybody hunting for new particles, two more were welcome. But, working back from the decay products, calculations revealed that the tau and theta had the same mass. Two diligent young Chinese researchers working in the US, Chen Ning ('Frank') Yang and Tsung-Dao Lee, thought it was bizarre for two apparently different particles to have the same mass, and suspected that the theta and the tau were somehow two different faces of the same particle, despite having different parities.

Both Yang and Lee had left their native China for Chicago in 1946. Yang had first worked with Enrico Fermi, while Lee studied astrophysics. From 1951 to 1953, the pair worked together in the shadow of Einstein at Princeton's Institute for Advanced Study. When Lee left

Princeton for New York's Columbia University, the pair continued their collaboration, puzzling over the new particles, suspecting that the theta–tau picture should not be taken at face value. But how could the same particle decay in two different ways? For this to be so, Lee and Yang had to throw overboard some underlying assumptions. When throwing something overboard to save a sinking ship, a heavy object is the most effective.

There were two apparently solid underlying assumptions about quantum behaviour: firstly, that it would not be basically altered by left–right mirror reflection – left-handed particles would reflect as right-handed, but otherwise things would happen in the same way and at the same rate; secondly, the behaviour would not be altered by a mirror that reflected particles as antiparticles and vice versa – viewed in this mirror the antiparticles would behave in the same way as particles. Lee and Yang re-examined the evidence for both mirror symmetries, which everybody had simply *assumed* were watertight. Lee and Yang boldly challenged these assumptions, showing that for particle decays this had never been proved conclusively.

At Columbia University, New York, the Chinese woman physicist Chien-Shiung Wu was an expert on beta decay, nuclear processes in which a component neutron decays into a proton. Beta decay, discovered by Rutherford in 1898, is one example of radioactivity, in which an unstable nucleus changes into another, more stable, one. Hearing Lee and Yang's suggestion, Wu proposed a new experiment using a form of cobalt that is beta radioactive. The electrically charged cobalt nuclei spin around their axes, behaving like tiny magnets. Wu's idea was to line these magnets up in a powerful magnetic field, like an array of compass needles. To ensure that the nuclear compasses would not jiggle out of line and would remain locked in the magnetic field, they would be frozen in liquid helium. The experiment therefore had an inbuilt directionality, and the idea was to see if this directionality would affect the emerging electrons. If reflection symmetry was OK and Lee and Yang's suspicion was wrong, as many beta decay electrons should spray out in the direction of the magnetic field – the spin orientation of the decaying nuclei – as in the reverse direction. Preparing the experiment took longer than actually doing it, and, on 29 December 1956, Lee got a telephone call from Wu. The electrons did not spray out uniformly, preferring to emerge pointing away from the magnetic field that lined up the clockwise-spinning nuclei. In January 1957 came the conclusive proof – when the direction of the magnetic field was reversed so that the cobalt nuclei lined up in the opposite direction, the emerging electrons switched direction too. Beta decay electrons are sensitive to direction.

This was not just a tiny correction. The startling 40 per cent asymmetry seen in Wu's experiment had been overlooked for half a century. Everybody had assumed that beta decay electrons were uniformly spread and nobody had bothered to check. On 15 January 1957, the front page of the *New York Times* carried the headline 'Basic concept in physics is reported upset in tests.' At the subatomic level, Nature was not ambidextrous. Parity was not conserved. Left could be differentiated from right by doing a physics experiment. In Alice's left–right mirror, beta decay 'smoke' would have to come down the chimney. Lee and Yang, aged 31 and 35 respectively, were awarded the Nobel Prize for Physics that same year, and their discovery became a physics landmark.

Lee and Yang had also suggested that the particle–antiparticle mirror could be flawed. To investigate this possibility, two experiments – by Richard Garwin, Leon Lederman and Marcel Weinrich at Columbia University, New York, and by Jerome Friedman and Val Telegdi at Chicago University – looked at multiple particle transformations in which a pion decays into a muon, which in its turn decays into an electron. The experiments found that the muon prefers to emerge spinning in a particular direction. For a positively charged pion, the muon's spin points backwards, against its direction of motion. The antiparticle of a positively charged pion is a negatively charged pion. In these decays, the muon emerges with its spin pointing in the direction of its motion. Looking in a mirror that changes particles to antiparticles, the anti-smoke comes down the chimney.

These broken mirrors shook the foundations of physics. Normally used to taking scrupulous care in their experiments, physicists were bewildered to find that not one, but two comfortable intellectual chairs had been pulled from under them. Wolfgang Pauli wrote 'Now the first shock is over and I begin to collect myself again . . . '.

With the foundations of their understanding severely damaged, nervous physicists looked for a new mirror that would not reveal any quantum flaw. Pauli himself helped come up with the answer. For the subnuclear world, the ordinary mirror above Alice's fireplace has to be replaced by an extended mirror that carries out three reflections simultaneously – switching particle to antiparticle and vice versa, changing left to right and vice versa, and reversing the arrow of time. Physicists call these three reflections respectively C (for charge), P (for parity) and T (for time). The CPT mirror changes Alice into a mirror-image Anti-Alice going backwards in time. The subnuclear world, reflected in this 'antimatter mirror', should look the same, said Pauli. CPT radioactivity would give out positrons rather than electrons and the antismoke would still go up the chimney.

THE STRANGE WORLD OF THE KAON

The tau–theta puzzle could at last be resolved. The tau and the theta, despite decaying in different ways with opposite parities, are one and the same particle. Called instead the K-meson, or kaon, it is electrically neutral, and carries a charge of one positive unit of strangeness. The corresponding kaon antiparticle, still electrically neutral, has to carry one negative strangeness charge. The tau–theta dilemma had been resolved, but was replaced by another conundrum: how could one particle be schizophrenic and have two different kinds of behaviour, decaying in different ways? If parity is not conserved, what is?

The kaon and its antiparticle are formed when a high-energy cosmic ray particle hits a nucleus in the atmosphere or when a high-energy particle beam from an accelerator crashes into a target. In these reactions, strangeness, like electric charge, has to be conserved, so as many kaons are produced as antikaons. However, strange particles, once introduced on the subnuclear scene, are not at ease and try to melt into the background – no sooner are they formed than they begin to decay, shedding their uncomfortable strange attire by decaying into other, non-strange, particles. Although carefully conserved during the process which created the kaons, strangeness is cast aside in the subsequent kaon decays. In this race to shed strangeness, the neutral kaon and its antiparticle, originally distinguished only by their opposite strangeness labels, look the same. The neutral kaon and its antiparticle both decay in the same way, so it is impossible to say by looking at the emerging particles whether they come from a particle or an antiparticle. How then can a neutral kaon beam be labelled at all?

Most of the time, a neutral kaon decays into two particles (pions) with even parity. But it can also decay into three, with odd parity. Decaying into three particles is more difficult than decaying into two as the decaying kaon has to labour to provide the extra mass. In any situation, having three offspring is more expensive than having two. Labelling a particle by its progeny is a sensible thing to do when the progeny is more noticeable than the parent. In the Arabic-speaking world, it is common for a child's name at birth to use the patronymic, to honour a parent through the word 'ibn', meaning 'son of'. However, if the child becomes distinguished, the father can opt to change a name to honour instead the child, adding 'abu', meaning 'father of'. Adopting this principle, a naturally mixed beam of neutral kaon particles and antiparticles can be labelled as a mixture of two varieties, one of which is destined to decay easily – into two particles, the other surviving for a hundred times longer before giving three daughter particles. Physicists speak of the short-lived and the long-lived neutral kaon.

The two- and three-pion decays have a deep significance. With particle–antiparticle reversal (C) and left–right symmetry (P) in disgrace, Pauli's CPT antimatter mirror would work if the violations of C and P symmetry somehow compensated for each other, so that a combined CP mirror – a simultaneous left–right and particle–antiparticle switch – gave an accurate reflection. The short- and long-lived versions of the neutral kaon have different properties under the combined CP operation, being respectively symmetric and antisymmetric. Two pions look very different to three pions. The neutral kaon was the natural testbed for the CP mirror. Rather than being labelled as parity even or parity odd, a neutral kaon instead should be even or odd under the combined CP operation. Was this new CP parity conserved?

The enigmatic kaon meanwhile had come up with more quantum surprises. If a neutral kaon beam is left alone, after a few metres all the short-lived variety disappears, leaving an apparently pure long-lived kaon beam, which decays into three pions. However, if the surviving kaon beam hits a metal target, the kaons are shaken up and the quark composition of the beam is altered. The neutral kaon beam, formerly free of the short-lived variety, once more becomes a mixture of the short-lived and long-lived varieties. As if by magic, short-lived kaons, decaying into two pions, reappear (Figure 8.2). This 'now you see it, now you don't' conjuring trick, called 'regeneration' in the physics trade, is not magic at all, but a dramatic demonstration of quantum mechanics at work. There is no other way of accounting for the reappearance of short-lived kaons after their time is up. As Richard Feynman said, 'It is not based on an elegant hocus-pocus. We have taken the superposition principle (of quantum mechanics) to its ultimate logical conclusion. It works.'

In 1963, this eerie regeneration of short-lived neutral kaons motivated James Cronin and Val Fitch, working at Brookhaven, to take a closer look. They set up a detector in what had been called 'Inner Mongolia', a difficult to access circle of land inside the Brookhaven accelerator ring. Because getting to this piece of land meant climbing over the ring, few people went there. Butterflies basked and wild orchids grew. Cronin and Fitch briefly disturbed the butterflies that summer. By October, they had finished all their measurements and started to analyse their data. Concentrating on the long-lived kaons at the far end of the beam, and with no regeneration material to shake up the quarks, they expected to see only the characteristic three-particle decay signature. But there was something else too. Afraid of the implications, they assumed it was a hiccup and waited for it to go away as more analysis

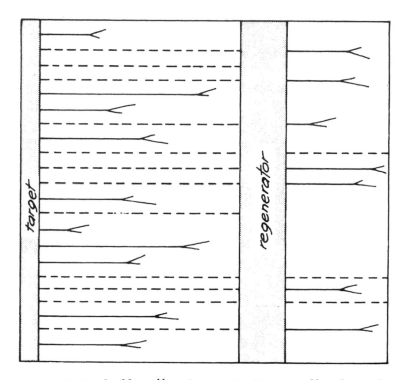

FIGURE 8.2 Neutral kaon 'regeneration'. In a neutral kaon beam, after a few metres all the short-lived variety disappears, leaving an apparently pure long-lived kaon beam, which decays into three pions. However if the surviving kaon beam hits a metal target, the kaons are shaken up and their quark composition is altered. The neutral kaon beam, formerly free of the short-lived variety, once more becomes a mixture of the short-lived and long-lived varieties.

piled up. But the unexplained effect was still there at the end of the year, and it was time to prepare papers for a major meeting in Washington in the spring of 1964. Still not sure about what they were seeing, Cronin and Fitch sent in a non-committal abstract. Their paper was returned as they had unwittingly violated a rule – all abstracts had to be just a single paragraph, but theirs had been two. Their paper did not feature at the Washington meeting in April 1964. For another six months, Cronin and Fitch sat on their unexplained result and tried to smother it, applying test after test. After trying everything they could think of, on 10 July they decided to go public. They had seen about one long-lived kaon in five hundred, instead of decaying into three pions, choosing instead the two-pion route. After Lee and Yang had shown that the simple particle–antiparticle and left–right mirrors could not be relied upon, physicists had hoped that the combined CP mirror would come to their aid. But, in the spooky world of the neutral kaons, this CP mirror too is flawed.

The neutral kaon and its antiparticle are only distinguishable by their opposite strangeness labels. The problem is that, once the neutral kaons have been formed, they are no longer sensitive to the strangeness that once distinguished between them. The neutral kaons are Nature's identical twins – initially distinguishable at birth, but causing confusion ever after. A neutral kaon, once formed, can forget its own strangeness and willingly be confused with its antiparticle, and vice versa. When the kaons finally decay, they reflect their assumed strangeness and not their birthright.

Because of CP violation, the decays of the neutral kaons provide a way of defining positive charge that is not ambiguous. The choice between which type of charge is positive and which negative cannot be decided merely by the spin of a coin – it is bizarre but true that the electrically neutral kaons say in which direction the coin of charge has to fall. CP violation also predetermines what is matter and what is antimatter. Before meeting a visitor from a distant galaxy, it would be important to know whether the visitor was composed of matter or was instead an antivisitor. As shaking hands with an antivisitor would result in total annihilation, a wise precaution would be to invite the visitor first to do a CP violation experiment with neutral kaons and to communicate the result. Only then could a decision be taken as to whether a meeting should be scheduled.

If the CPT antimatter mirror is to remain accurate, then the mirror of time has to be broken in sympathy with the mirror of CP. Neutral kaons are somehow sensitive to the arrow of time, a one-way street in the microworld. Subsequent experiments have compared the kaons' strangeness birthright with their apparent strangeness when they decayed, tracking how neutral kaons change into their antiparticles and vice versa. Comparing the rate at which kaons change into antikaons with that of antikaons changing into kaons shows that the mirror of time is skewed, and in such a way as to compensate the violation of CP symmetry. The two effects cancel out, assuring that CPT reigns.

In everyday situations, the arrow of time has an obvious direction. It is easy to see when a film is being run backwards – divers defy gravity, water splashes are replaced by a smooth pool, and fragments reassemble into complete objects. On a longer reversed time-scale, adults become children and death becomes life. Life is a continual battle against disorder, things wearing out or going wrong. On the cosmic scale, galaxies continue to fly apart in the wake of the Big Bang – the Universe gets larger. Here too, reversing the arrow of time would produce an unfamiliar picture. But, in fundamental theory, time is just another variable. A lone hydrogen atom does not wear out. A film of the atom run backwards would not show anything unusual.

The outcome of a particle situation has to take account of all possible outcomes, even those still in the future. In this way, Feynman had introduced the idea that antiparticles are particles travelling backwards in time. Feynman accomplished this time travel by insisting that his equations worked equally well no matter which direction the arrow of time pointed. Whatever else was being overthrown, on the subatomic scale the mirror of time appeared fundamental, inviolable. However, a few times in every thousand, running the 'videotape' of a neutral kaon interaction backwards in Nature's VCR does not return it to the point of departure. The neutral kaons put a valve in the passage of time, so that some events only happen in one direction. The neutral kaons are composed of quarks, and new results showed that these quark arrangements mature with time in a way that is not time-reversible. The quarks in the neutral kaons 'show their age'.

Is this obscure link between the bizarre world of neutral kaons and the direction of the arrow of time just an accident, or does it signal something deeper? Cronin and Fitch had to wait until 1980, sixteen years after their CP-violation discovery, before receiving their Nobel prize, but elsewhere the implications of their result had been noticed almost immediately.

THE CONSCIENCE OF MANKIND

In 1965, the gifted Russian scientist Andrei Sakharov had turned his attention to cosmology, understanding the origins of the Universe, and why the Universe is as complex as it is. Sakharov seized on the new result of CP mirror asymmetry, one of the most subtle differences imaginable between the behaviour of matter and antimatter on the smallest scale imaginable, and intellectually amplified it to suggest an explanation for the biggest problem in cosmology – to understand why the Universe apparently contains only matter and no antimatter. Perhaps, said Sakharov, the bizarre properties of the neutral kaon held the key to the Universe.

Sakharov's father was the author of a famous Russian physics textbook. The young Sakharov was taught at home, but his career at Moscow University was interrupted by the Second World War, and he became an engineer at a munitions factory at Ulyanovsk, on the Volga River. Here he had his first encounter with physical hardship. Resuming his scientific studies in 1945 at the Lebedev Institute of the Societ Academy of Sciences in Moscow, Sakharov impressed the influential Igor Tamm. Tamm, with Ilya Frank, went on to share the Nobel Prize for Physics with Pavel Cherenkov in 1958, but in the late 1940s was busy with the Soviet nuclear weapons programme. Under Tamm's wing, Sakharov went on to play a vital role in the race to build the Soviet hydrogen bomb. The first Soviet H-bomb was exploded in 1953, less than a year after the first American H-bomb in 1952, despite work having started much later. Sakharov was showered with honours, Hero of Socialist Labour, the Stalin Prize and the Lenin Prize, and was elected to the prestigious Soviet Academy of Sciences at the age of thirty-two, one of the youngest ever to achieve this honour.

Later, Sakharov recalled the oppressive secrecy of those years working on the bomb. All notes and calculations had to be done on special notebooks with numbered pages. Notebooks that were still in use were locked up at the end of the day for safe-keeping, and when notebook pages were no longer needed, they were burnt and their destruction logged. One such numbered note was sent from the secret bomb group to the neighbouring applied mathematics institute with a request for a calculation. When the calculation was complete and the results passed back to the bomb group, a secretary duly burned the original request note, but forgot to log its destruction. After a time, the apparent non-destruction of one page of a notebook was noticed, and the local head of security, trying valiantly to cope with all this destructive bookkeeping, had an ominous top-level visit. Such visits rarely meant promotion. After apparently spending a normal weekend with his family, the bomb group's head of security arrived at work early the following Monday morning and shot himself. An assistant spent more than a year in prison. Nevertheless, Sakharov thrived, and was happy to be able to work for his country. After his hydrogen bomb work, he switched his attention to the containment of thermonuclear power, and helped invent the idea of containing the thermonuclear fuel in a doughnut-shaped magnetic 'bottle', later to become known all over the world as the tokamak.

His duty to the Soviet atomic weapons and energy effort discharged, he began to feel uneasy about the arms race and was influential in getting the Soviet government to sign the Limited Test Ban Treaty in 1963. Now Sakharov felt he could indulge in some heady physics specula-tion. Fitch and Cronin's result had stimulated his interest in the conundrum that a Universe created in a Big Bang which must have pro-duced equal amounts of matter and antimatter now appears to contain only matter. Where has all the primordial antimatter gone? Equally perplexing to Sakharov was why there was so much more radiation in the Universe than matter. An 'average' cubic metre of Universe con-tains just one proton but a billion quanta of radiation.

Until 1924, Einstein's equations of general relativity had never been solved satisfactorily. Then the Russian mathematician Alexander Friedmann showed how the equations predicted a continually expanding Universe. If the Universe is expanding, it cannot have been expanding for ever. Sometime in the past it must have started out from a tiny spark which ignited a primordial explosion, the Big Bang. Subnuclear particles were cooked in the fiery aftermath of this Big Bang, with extremely powerful radiation creating all sorts of particle–antiparticle pairs. As the Universe cooled, less and less particle–antiparticle pairs could be formed, while those already in existence could annihilate with each other to produce puffs of radiation. On this simple picture, at any stage there should be as many particles in the Universe as antiparticles.

Sakharov looked wryly at the composition of an average cubic metre of Universe. It had a billion quanta of radiation, one proton and no antiprotons. Tracking back to just after the Big Bang, the same cubic metre should have had a billion quanta of radiation, a billion antiprotons, and a billion and one protons. Sakharov could not accept these numbers. Why the odd proton? Where had the antimatter gone? Sakharov realized that distant galaxies in principle could be made of antimatter, but knew that no serious antimatter had been found in cosmic ray messenger particles from the depths of space. More important, he could not see how antimatter and matter could have become so separated. For Sakharov, antimatter had slipped off the map of the Universe and he wanted to understand why.

For antimatter to have become extinct, Sakharov put forward a three-point explanation. First, the Big Bang explosion must at some time have been so cataclysmic that particle–antiparticle creation briefly got out of hand, more pairs being created than were reabsorbed back into radiation. Cosmologists now know this must have happened – the present Universe is much larger than a sphere of light rays which started out from the Big Bang spark. Sometime in its past, the Universe must have expanded faster than light itself. Most of the Universe we have not yet seen, its light, despite travelling at 300,000 kilometres per second over the ten billion years or so of the Universe's existence, not yet having had time to reach us. Not only is the Universe very large, but in the depths of outer space between galaxies it is very uniform. How do distant parts of the Universe, further apart than can be connected by any Big Bang light ray, 'know' they have to look the same? In the first fraction of a second after the Big Bang, the Universe must have 'inflated' faster than the speed of light and particle–antiparticle pairs were produced faster than they could be reabsorbed.

Secondly, realized Sakharov, some mechanism had to tilt the balance in favour of matter. With Cronin and Fitch's discovery and its implications for the arrow of time, Sakharov thought he had found the answer. But was the tiny subnuclear effect first seen at the Brookhaven laboratory in 1964 enough to explain the apparent absence of antimatter in the whole Universe? Probably not, say physicists. But other quarks could come to the rescue. Heavier quarks, more exotic than strangeness, could show larger effects. Making B particles containing the 'beauty' (sometimes known as 'bottom') quark and manufacturing enough of them to probe the arrow of time has become a major focus of today's particle physics research.

Sakharov's final criterion for a matter-dominated Universe was perhaps the most difficult to swallow. The proton itself, the bedrock particle of the Universe, has to be slightly unstable, he said. Sitting still, the quark-filled proton would have to disintegrate into electrons and other light particles. How could this be so? The very existence of a Universe filled with nuclei was incompatible with such an idea. But Sakharov pointed out that the level of proton instability needed was so small as to be almost undetectable. The proton instability he was looking for implied that if all the protons that have decayed over the entire fifteen billion year history of the Universe were put together and compared with the rest of the Universe, it would be like comparing a crumb a quarter of a millimetre across to one of Saturn's moons! But experiments are trying to capture this effect; the plan being that if enough protons are looked at for long enough, one of them will decide to decay, leaving a characteristic subatomic fingerprint.

As he continued to work on these ideas, Sakharov's conscience troubled him. In the mid-1960s he became increasingly critical of corruption and power in the Soviet system, pointing to how unilateral Kremlin decisions had led to a retarded technology, and had resulted in

FIGURE 8.3 Andrei Sakharov, the martyr of *perestroika*, who first saw how a Universe composed of matter could have evolved from one that was initially matter–antimatter symmetric (photo CERN).

widespread pollution and environmental damage. Censorship and red-tape were everywhere, as was alcoholism. Sakharov began what seemed to be a fearless, almost foolhardy, campaign as an activist, and in 1974 went on the first of four serious hunger strikes to underline his message about human rights in the Soviet Union. In his 1975 Nobel Peace Prize citation, he was described as 'spokesman for the con-science of mankind', but, with no exit visa granted, was not able to go to Oslo to collect his award. His criticism of the Soviet invasion of Afghanistan led to him being exiled in Gorki in 1979, where he mounted a further series of epic hunger strikes. But, in 1986, with the advent of *perestroika*, he emerged as a figurehead of the new move-ment and a popular champion, a living example of the indomitability of the human spirit. But years of hunger strikes and poor treatment had taken their toll of his health, and he died on 14 December 1989, the saint and martyr of *perestroika*. His legacy is one of the most intellectually challenging theories of physics – tiny asymmetries in the way some subnuclear particles are reflected in the mirror of time could explain why the Universe is the way it is. These small asymmetries could have provided a microscopic loophole in passing through which a whole Universe initially composed of equal amounts of matter and antimatter was transformed into one composed entirely of matter.

If this is a key to the Universe, these asymmetries are so difficult to measure that it is hard to fashion the key with enough precision to fit the lock. At the dawn of a new millenium, new physics machines are setting out to mass produce B particles, containing heavier quarks than those in the kaon, to make new measurements of CP violation where the mirror of time might reveal larger such asymmetries and provide a key which turns more easily.

9   The cosmic cork-screw

In September 1956, a bespectacled 30-year-old Pakistani physicist was returning to Cambridge, UK, from a physics meeting in Seattle. Instead of taking a scheduled flight, Abdus Salam boarded a US Air Force flight destined for an air base in England. In those days the US Air Force generously supported scientific research in European universities, a heritage of the immediate post-war research glut when the US military was a major paymaster of science. One of the perks was that physicists working in Europe could have transatlantic travel free of charge via special MATS – Military Air Transport Service – flights for US servicemen and their families.

Although the European scientists were glad of this travel opportunity, these flights were notoriously inconvenient and uncomfortable. Check-in meant reporting to a US Air Force base, and the UK terminal at Mildenhall in Suffolk was a long way from anywhere. Instead of tickets, there were multiple copies of 'flying orders' which had to be successively surrendered at various stages of the journey. The planes were propeller-driven and laboriously slow, some fifteen hours for a trip from Mildenhall to McGuire Air Force Base, New Jersey. There were no in-flight movies and the flights were frequently full of families with young children, noisy and excited about the prospect of going home or moving to a new country. The overnight west–east trip was particularly uncomfortable.

At the Seattle physics meeting, Salam had listened to Frank Yang explain his and Lee's suspicion that subnuclear mirror symmetry could be broken. The historic experiments that were to prove the ideas right had not yet begun, but Salam, receptive to innovative ideas, was prepared to believe what Lee and Yang were suggesting. As the US Air Force flight droned eastwards through the night, Salam wondered what could be special about nuclear decay that would make it mirror sensitive.

THIEVED ENERGY

In the early years of the twentieth century, physicists found that nuclei decayed in three different ways, which were termed, naturally enough, alpha, beta and gamma radioactivity, depending on what the decays produced. Alpha radioactivity produced helium nuclei – alpha particles; beta decay produced electrons; and the gamma variety gave just radiation. (In 1939, a new form of decay had to be added to this list – nuclear fission.)

Thirty years before the parity episode, experiments studying nuclear beta decay had come to a baffling impasse. The amount of material emerging in these decays was less than the energy of the original nucleus. In addition, the individual rotations – spins – of the initial nucleus and the emerging particles did not match. If a rotating nucleus broke apart, the separate rotations of the fragments were expected to add up to the rotation of the parent nucleus. Either something invisible was emerging from the decays, or material and rotation just disappeared – a fine distinction. Physicists were reluctant to abandon the golden rule of energy conservation – in all physics processes, income and expenditure of energy had to exactly balance. But, at a time when the new quantum ideas were challenging many cherished principles, Niels Bohr, the architect of atomic electron orbits, had spoken in public of being prepared to wave goodbye to strict energy accounting at the quantum scale. Perhaps, suggested Bohr, in the quantum world energy could just disappear without trace.

Energy had been considered a hard currency in subnuclear physics. Could these reserves just evaporate? In 1931, Wolfgang Pauli, rarely at a loss for an unconventional idea, hesitantly put forward a solution to this energy crisis, calling it a 'desperate remedy'. Speaking at a meeting in Pasadena, California, Pauli suggested that beta decay, as well as giving an electron, also produced an additional particle that carried off energy but could not be detected. This invisible particle had no mass of its own, only kinetic energy because of its motion. It also rotated like an electron. The only way to 'see' these particles, he said, was by comparing what came out of the decay to what was there originally. Even the intellectually courageous Pauli felt nervous about such a far-fetched suggestion and did not provide any written version of his speech, but his remarks were picked up by the ever-alert scientific antennae of the *New York Times*.

Although this was the first time Pauli had voiced his idea publicly, it had been mentioned before. Invited to attend a conference on radio-activity in Tübingen, Germany, in December 1930, Pauli had preferred to stay in Zurich and attend a pre-Christmas ball. In December 1929, Pauli had married a dancer, Käthe Deppne, but the marriage was on the rocks from the start, the dancer preferring the company of a chemist, Paul Goldfinger, to Pauli. Divorce came in November 1930, when Pauli wrote 'had she taken a bullfighter, I would have understood, but an ordinary chemist...'.

Pauli was on the loose and anxious to go to the dance. In his letter of apology for not attending the physics meeting, he suggested to the 'Dear Radioactive Ladies and Gentlemen' gathered at Tübingen that their beta decay problem might be solved by resorting to invisible particles which carried off rotation and energy. These ghost particles Pauli proposed calling neutrons, and for a few years there was confusion between these invisible lightweight Pauli neutrons and the electrically neutral particle proposed by Rutherford as a constituent of nuclei. When the nuclear particle was discovered at Cambridge in 1932, this naturally had first claim on the neutron title, and another name had to be sought for Pauli's hypothetical particle. Enrico Fermi in Rome took up the challenge, calling the Pauli particle the 'neutrino' – 'neutral little one' in Italian.

The neutrino's job was to carry off energy without being seen, a sort of subnuclear energy thief. To remain invisible it had to avoid being seen by any other particle in its path. But even well-organized thieves run a risk. What were the limits of the neutrino's invisibility? In 1934, Hans Bethe and Rudolf Peierls in Germany calculated that a lone neutrino could traverse an imaginary ocean of almost intergalactic proportions before being absorbed in a nucleus. Contemplating their awesome answer, Bethe and Peierls concluded that the neutrino was undetectable other than through subnuclear energy accounting. Add up what was there at the beginning and what was there at the end, and any unaccounted difference had to be attributed to neutrino robbery. Like a subnuclear Robin Hood, the neutrino redistributed subnuclear wealth without being apprehended. Within a few years, Bethe and Peierls left Germany to begin new careers, Bethe in the United States and Peierls in the United Kingdom. Both were to be intimately linked with the development and exploitation of nuclear energy, but, in 1934, neither Bethe nor Peierls realized that nuclear energy had a future and that nuclear reactors would one day be built. And they certainly had no idea of the vast torrent of neutrinos these reactors would produce.

When Bethe and Peierls had calculated that a neutrino would have to traverse the equivalent of an intergalactic ocean (1,000 light-years of water) before being caught, they meant that such an immense obstacle would provide an even chance of one neutrino being absorbed. On the other hand, if the swimming-pool were of more modest terrestrial dimensions, the probability of capturing the neutrino would be reduced a billion billion times. If the Earth had just two inhabitants, the probability of them seeing each other at any time would be very small. With ten billion inhabitants, it is the probability of not meeting someone that becomes small. For neutrinos, the one in a billion billion chance of a neutrino being snared in a swimming-pool is negligible. But, if many billions of neutrinos were to become available, then the chance of a few neutrinos being absorbed would no longer be negligible.

SEEING THE INVISIBLE

In 1945, a powerful and unexpected new source of neutrinos burst on the scene. The atomic fission bomb unleashed more of Pauli's ghost particles than anything ever encountered before on Earth. But that was not the reason why it had been built. The Los Alamos laboratory had been constructed during the Second World War by the US military in record time, the objective being to develop weapons of mass destruction. With the sinister task accomplished in 1945, many Los Alamos scientists, including Bethe and Peierls, migrated back to their universities, but Los Alamos continued its weapons development programme. As an occasional respite from their grim task, Los Alamos researchers were encouraged to think of other physics problems. In 1951, two young researchers, Fred Reines and Clyde Cowan, were looking for such intellectual recreation. Perhaps, they mused, enough neutrinos would be released by an atomic bomb explosion for a few to become visible. The pair set about designing a detector which could survive a nearby atomic bomb explosion but still be sensitive enough to register feeble neutrino hits. They initially toyed with putting their detector in a vacuum chamber to protect it against the bomb's shock wave. But this proved too difficult. Such an experiment would last only a fraction of a second. If anything went wrong, the experiment would have to wait for the next bomb. Reines and Cowan turned their attention instead to siting the detector at a nuclear reactor. A reactor gave out neutrinos more slowly than a bomb explosion, but, on the other hand, the detector could just sit there and wait . . . Perhaps one reactor neutrino might occasionally hit a proton in the detector, producing a neutron and a positron. This anti-electron would immediately annihilate in the apparatus, giving a flash that could be recorded.

In 1953, Reines and Cowan's prototype 300-litre neutrino target at the reactor at Hanford, Washington, showed a glimmer of a signal. But the count rate was too feeble. There were not enough reactor neutrinos to multiply the faint collision chances and give a reliable signal. To clinch their claim, Reines and Cowan moved to a bigger reactor at Savannah River, South Carolina, and used five times the amount of neutrino target. Over 100 days, they clocked three neutrino counts per hour. Knowing how many neutrinos the reactor was pumping out, this count rate agreed with the calculation that Bethe and Peierls had done twenty years before. Reines and Cowan's ingenuity and patience, combined with the flood of neutrinos from the reactor, meant that Pauli's neutrino was for real. But it was no longer invisible. On 14 June 1956, the American physicists telegrammed Pauli in Zurich – 'We are happy to inform you that we have definitely detected neutrinos.' It had taken 25 years for Pauli's prediction to be confirmed, however it took almost 40 years before Reines was summoned to Stockholm for the 1995 Nobel award. Unfortunately Clyde Cowan died in 1974.

THE LEFT-HANDED PARTICLE

The newly discovered neutrinos, as well as the idea of broken mirror symmetry, were very much on Abdus Salam's mind that airborne September night in 1956. 'I could not sleep', he recalled. 'I kept reflecting why Nature should violate left–right symmetry . . . Now the hallmark of most interactions was the involvement in radioactivity phenomena of Pauli's neutrino. While crossing the Atlantic, there came back to me a deeply perceptive question about the neutrino which Rudolf Peierls had asked me in an examination a few years before – "Why is the neutrino's mass zero?"'

In 1949, Salam had arrived as a young research student at Cambridge with little knowledge of the new quantum electrodynamics ideas of Feynman and Schwinger. Absorbing these ideas in record time, he had gone on to apply them to other particles. With several landmark physics papers to his name, in 1951 Salam returned to Pakistan to become, at the age of 25, professor at the College and University of Punjab, Lahore. Despite the local prestige of his new position, Salam found himself cut off from the news, excitement and continual stimulation of modern research, and there was no up-to-date library. He realized this excitement was his life-blood, and in 1954 he left his native land to return to Cambridge, this time as a university lecturer.

Salam, like Feynman, was blessed with piercing insight which could lock on to the most intractable of problems. Thus, when Peierls had asked Salam why the neutrino's mass was zero, Salam knew well that Peierls had not been testing him, but instead had been playfully asking a question to which nobody, not even Peierls, knew the answer. But, during that comfortless night in the air, an answer came. In his cramped seat, Salam wrote down a prototype neutrino equation. It looked something like Dirac's electron equation, which gave particles and antiparticles whose spins pointed either up or down, but Salam's equation had a different arrangement of the four-dimensional Dirac matrices. Remembering Peierls' question about the neutrino having no mass, Salam dropped the mass term from his equation. Immediately he saw that the matrices acted like a rotation switch – the neutrino could spin one way but not the other. Salam realized that a zero-mass neutrino could be a miniscule cork-screw, drilling its way through space at the speed of light. Travelling at this speed, no other particle could overtake it, so there was no other vantage-point from which to view the neutrino's spin. It would always appear to point the same way. Like Dirac's equation which delivered an electron and a positron solution, so Salam's gave a neutrino and an antineutrino. These rotated in mutually opposite directions, one clockwise, the other anticlockwise. The only ambiguity was which way the neutrino and the antineutrino rotated – was the neutrino rotating clockwise and the antineutrino anticlockwise, or were the rotation assignments the other way round? The equation could be written either way. Whatever the answer, it mattered little to Salam that night. A conventional right-handed cork-screw looks left-handed when reflected in a mirror, so a neutrino reflected in a mirror would no longer look like a neutrino. It would instead look like an antineutrino. Salam had realized that the neutrino, Pauli's ghost particle, was the culprit which broke the mirror of radioactive decay.

The following morning, an elated Salam bustled off the plane and rushed as fast as he could to his small office at Cambridge, where he consulted some books and calculated a few consequences of his new theory. Even more elated by the way everything seemed to be working out, he rushed on to a train to Birmingham, where Peierls lived, to tell the famous physicist that he now had the answer to the trick question posed a few years before. Peierls was surprised to find Salam on his doorstep, but listened to what he had to say. His reply was typically kind but firm. 'I do not believe left–right symmetry is violated at all.' With Mrs Wu still assembling her epic experiment at Columbia

University, Salam had knocked too early on Peierls' door. But the young Salam was insistent, and gave his neutrino paper to a physicist who was going to visit Pauli, the father of the neutrino, in Zurich. The reply soon came: 'Give my regards to my friend Salam and tell him to think of something better.'

Quashed, Salam hesitated before submitting his massless neutrino idea for publication. Four months later, on 24 January 1957, Pauli wrote again to Salam. Wu's result on the left–right asymmetry in cobalt decay had been published, Pauli had changed his mind and Salam's ideas had been vindicated. Meanwhile, Lee and Yang in the United States and Lev Landau in Russia had arrived at a similar conclusion about the neutrino and its mirror reflection. However, Pauli still had reservations – 'For some time, I looked at this particular model with a certain scepticism, since it seemed to me that the special role of the neutrino was emphasized too strongly.'

But physicists still had to find out which way the neutrino and the antineutrino pointed. The massless neutrino theory was like Alice waking from a dream, not being able to tell whether she was in the mirror world or back on the 'right' side of the fireplace. The theory could only say that the neutrino and antineutrino cork-screwed their way through space at the speed of light in opposite directions. Only an experiment could determine which way the cork-screw turned – was it left for neutrino, right for antineutrino, or vice versa? The fact that the neutrino had been detected at all was an experimental triumph which had surprised many physicists. Doing another experiment to fix the direction of the neutrino cork-screw was a major challenge.

Maurice Goldhaber had begun his physics studies at the Hebrew University in Jerusalem. After moving to Berlin in the 1930s, he quickly had to pack his bags again, this time moving to Cambridge, where under Rutherford's guidance he carried out pioneer experiments on the neutron. In 1938, he moved again, this time to the United States. After hearing of the neutrino discovery and the rotation dilemma, he mounted a phenomenally delicate experiment in which a nucleus absorbed an electron, one nuclear proton changing into a neutron and emitting a neutrino. This neutrino spins off, and the remaining nucleus should recoil backwards, but spinning the same way as the neutrino. After repeated careful measurements, they announced their result – the recoiling nuclei were left-handed, so neutrinos were left-handed. It was their antineutrino counterparts that were right-handed. The process Goldhaber had watched was:

$$\text{electron} + \text{proton} \rightarrow \text{neutron} + \text{neutrino}$$

Shuffling this reaction around, a particle moves from one side of the arrow of time to the other, becoming an antiparticle:

$$\text{antineutrino} + \text{proton} \rightarrow \text{neutron} + \text{positron}$$

This is the reaction Reines and Cowan had detected at Savannah River in 1956. The flood of particles from the reactor had, in fact, been antineutrinos. For neutrinos, the antiparticle had been discovered before the particle!

THEORY AND TRAGEDY

Machines can be built using left-handed screws just as well as right-handed ones. It is only convention which makes us tighten things by turning clockwise and loosen them by turning the other way. However, Nature has provided us only with a left-handed neutrino. If we complain and ask for a right-handed one, we have to resort to another kit of parts, antimatter. But this kit of parts has to be kept carefully distinct. Accidentally introducing an antimatter screw in a matter machine would have disastrous consequences!

Carrying no electric charge and being almost invisible, the neutrino is a very subtle particle, doing nothing for years at a time, despite flashing through matter at the speed of light. The difference between a neutrino and an antineutrino is even more subtle – whether it spins counter-clockwise or clockwise – as it does almost nothing.

In 1936, a chronically shy young Italian physicist called Ettore Majorana had an alternative neutrino idea. Majorana had worked with Fermi in Rome and had gone abroad to work with Niels Bohr in Copenhagen, then with Werner Heisenberg in Leipzig. Returning to Italy, he had to find a permanent university job, which in Italy is traditionally done by open competition. For this, he focused his intellectual powers and wrote a treatise 'A Symmetrical Theory of Electrons and Positrons', which called for an accompanying neutral particle which is its own antiparticle. The neutrino and the antineutrino, suggested Majorana, are one and the same, but with two spin possibilities: clockwise or anti-clockwise. For this to happen, the neutrino would have to have mass, and therefore can be overtaken by a 'mirror' travelling at the speed of light. In this mirror, the neutrino would appear to spin the other way. Other electrically neutral particles, notably the neutral pi meson, or pion, are their own antiparticles. Why not the neutrino?

The theory enabled the 31-year-old physicist to become Professor of Theoretical Physics at Naples Regia University in 1937. Research work needs stimulation, but, in Naples, Majorana had nobody to talk to. For the first time he also had to give lectures and talk to students, but this was not the kind of contact he sought. On the contrary, he found it very difficult. He remained in his office for long periods, and when he did emerge he would never walk in the centre of any corridor, always keeping close to walls. On 25 March 1938, the tormented physicist telegrammed the head of his department: 'I have taken a deci-sion which has become inevitable. There is not one iota of egoism in this, but I am conscious of the trouble that my unexpected disappear-ance will cause you and the students ... I also ask to be remembered to those who I have come to know and appreciate ... of whom I shall keep fond memories, at least until 11 o'clock tonight and possibly even after.'

A Sicilian, Majorana boarded the Naples–Palermo ferry. The next morning, in Palermo, he again cabled his university: 'I hope you have received the telegram. The sea has refused me and I shall return to Naples . . . I still intend to give up teaching.' The next day the ferry returned to Naples, but with no Majorana aboard. He was never seen or heard from again, despite a massive search, the offer of a reward and the intervention of Mussolini. Majorana had clearly decided to commit suicide by drowning himself, but had not been able to summon up the courage on the first trip.

His disappearance was a major blow to Italian physics. Another was soon to come. The same year, Enrico Fermi, whose family was part Jewish, was awarded the Nobel Prize for Physics. After the award ceremony in Stockholm, instead of returning to Italy, Fermi continued to the United States and remained there for the rest of his working life. He died of cancer in 1954, aged 53.

Majorana's memorial is his idea that the neutrino and its antiparticle are one and the same – physicists talk of a 'Majorana particle'. There is one acid test . If the hypothesis is correct, two nuclear beta decay processes could couple back-to-back and produce 'neutrinoless double beta decay', in which the nuclear charge changes by two units (instead of one in ordinary beta decay), but no neutrino is emitted – the neutrino from the first beta decay catalyses the second. The process would be extremely difficult to detect, but physicists have been patiently searching and new studies are still in the pipeline. Although Majorana has probably been dead for more than 40 years, his imaginative ideas live on.

Later, new bold ideas proposed that the neutrino, predicted by Salam *et al.* to have no mass at all, could in fact have a tiny mass, about the same fraction of an electron's mass as an electron's mass is of that of an atom. Some evidence is there, but to confirm these ideas requires huge experiments to provide data over years. Neutrinos do not yield their secrets easily. Whatever the outcome, Salam's massless neutrino is at least a very good approximation to a real one.

Just after formulating this theory, at 31, the same age as Majorana when he moved to Naples, Salam became Professor of Theoretical Physics at London's prestigious Imperial College of Science and Technology. As well as his physics research, Salam worked tirelessly to further the cause of science in developing countries. Remembering vividly his own isolation when he had returned to his home country in 1951, in 1964 he founded the International Centre for Theoretical Physics in Trieste, Italy, now a world-class research centre, where

FIGURE 9.1 Abdus Salam at Stockholm in 1979 for his Nobel prize, the first Pakistani to receive the coveted award (photo ICTP Trieste).

promising young scientists from all over the world can get a taste of front-line research early in their careers. The pinnacle of Salam's physics career came in 1979 when he shared the Nobel Prize for Physics with the US physicists Sheldon Glashow and Steven Weinberg for their unification of electromagnetism and the forces at work in neutrino interactions, called by physicists the 'weak force'. The culmination of a quarter of a century of work after the first steps had been taken by Enrico Fermi, this synthesis of two very different aspects of Nature was the twentieth-century equivalent of Maxwell's synthesis of electricity and magnetism into a single agency – electromagnetism. The final step by Weinberg and by Salam, working independently, came in 1967, almost exactly one hundred years after Maxwell first published his famous equations. In the 'electroweak' theory, a word invented by Salam in 1978, such different phenomena as a flash of lightning, the directionality of a compass, and nuclear beta decay could be understood as different manifestations of the same underlying force.

Salam was showered with honours from all over the world. In his own country, under the powerful rule of Ayoub Khan from 1958 to 1969, Salam wielded considerable influence. But, despite being the only Pakistani to have won a Nobel prize, Salam's status in his own country quickly became precarious. As a member of the minority Ahmadis Islamic sect, in 1974 he resigned his influential position as chief national scientific adviser when Pakistan's National Assembly under Zulfikar Ali Bhutto excommunicated the Ahmadis from Islam.

Also known as the Mirzais or Qadianis, the Ahmadis maintain that Mirza Ahmad, born in Qadian in northern India in the late nineteenth century, was the Mahdi, or Messiah, a view sacrilegious to conventional Islam. Mainly confined to Pakistan, India and East Africa, the small but visible Ahmadis sect is a frequent target of intolerance and discrimination, from both the religious orthodoxy and the mass of people. In 1979, after the announcement of his Nobel award, Salam was invited to Pakistan by President Zia Ul-Haq. He was scheduled to give a lecture on his work at Quaid-i-Azam University, Islamabad, but threats by a student group renowned for violence led to the talk being cancelled. Benazir Bhutto, during her first mandate as Prime Minister of Pakistan, refused to receive Salam. The ultimate rebuff came at a meeting at Salam's alma mater, Lahore's Government College, when a list of distinguished alumni was read out. Except one.

Born in a modest home in a small town in British India in 1926, Salam's supreme ability, inexhaustable energy and burning ambition enabled him to overcome the most intractable intellectual and political problems and brought international prestige. Few other sons of the Indian subcontinent achieved so much. Instead of being able to enjoy the fruits of his labours, by some cruel irony of fate, at the beginning of the 1990s his vital strength began to wane. At first he struggled valiantly to continue his scientific and administrative work, but on several occasions he was injured in falls. He was the victim of a rare neurological disorder which gradually destroyed his life force and his ability to communicate. He was told he had only a few years to live. At his research centre in Trieste, he could no longer function as its Director. As a tribute to its founder which he would be still able to appreciate before his powers waned completely, in 1994 the centre organized a three-day physics meeting which was attended by colleagues, admirers and former students from all over the world. One was Frank Yang, whose talk on mirror symmetry in 1956 had so much impressed the young Salam. The culmination of the meeting was to award Salam an honorary degree of the University of St Petersburg, Russia. The rector of the University made the trip specially. Salam listened from his wheelchair but could not speak. After the formal ceremony, participants stood patiently in line to offer Salam their congratulations. After famous professors, it was the turn of younger students. One of the last was a nervous young man from Pakistan, a young researcher who had succeeded in gaining one of the highly prized scholarships to Salam's centre. As he bent towards Salam in his wheelchair, he said 'Sir, I am a student from Pakistan. We are very proud of you.' Salam's shoulders shook and tears ran down his face.

A complete invalid, Salam retired to Oxford to be looked after by his wife, distinguished Oxford molecular biologist Louise Johnson. There were occasional visitors, but communication was extremely difficult except for those who could address him in his native language, Punjabi. Salam died in November 1996, but his contributions to physics and his thriving International Centre in Trieste remain his monument.

Antiparticle collision course

In March 1960, Bruno Touschek, a flamboyant Austrian physicist working at the Italian nuclear physics laboratory at Frascati, near Rome, had an idea which went on to transform antiparticles from a laboratory curiosity to a front-line research tool. With his physics background, Touschek had been drawn into radar and electronics work in Germany during the Second World War. In 1945, with Allied armies on German soil, the Gestapo was very nervous. With his habit of reading foreign-language newspapers, plus the fact that his family had some Jewish blood, Touschek was arrested in 1945. With the British army approaching, prisoners were force-marched from the Hamburg prison to another in Kiel. Touschek was taken ill and collapsed by the side of the road. An accompanying SS soldier raised his pistol and fired. Seeing a lot of blood, the soldier left the inert body where it was and the column moved on. However, Touschek suffered only a flesh wound in the ear.

During his wartime radar work in Hamburg, one of Touschek's colleagues had been a talented Norwegian engineer called Rolf Wideröe. As a research student in Karlsruhe in 1924, Wideröe had proposed an ingenious idea for a 'beam transformer'. In ordinary electromagnetic induction, a magnetic field makes a current-carrying conductor move at right angles to the direction of the field. Wideröe's idea was to dispense with the wire, and use a powerful magnetic field to act on the source of the current, the electrons, in a vacuum. In principle the electrons should whirl around the direction of the magnetic field, gaining energy, a transformer without any wires. But the device needed a much better vacuum than Wideröe could get, and his fertile imagination turned to other ways of making electrons go faster.

In 1930, the young American physicist Ernest Lawrence was

browsing in the library of the University of California at Berkeley, near San Francisco, and came across a paper by Wideröe. Lawrence could not understand German and so could not follow Wideröe's explanation of the difficulties of making a circular machine to accelerate electrons and other particles. Lawrence got the gist of the idea from the diagrams and went on to make it work. With his invention, the cyclotron, physicists were able to accelerate particles to high energies and open up new physics horizons (see chapter 7). In 1939, Lawrence was awarded the Nobel Prize for Physics.

In wartime Hamburg, Touschek had learned about these particle accelerators from Wideröe, who avidly followed the progress of his brainchild. In these magnetic racetracks, subatomic particles whirled round and round, with high-frequency electric oscillations acting as the 'hare' which goaded the subatomic greyhounds to higher velocities. Normally, these subatomic greyhounds were either protons or electrons. Could their antiparticles also be accelerated in the same way? Could the waiting detectors capture the moment of truth when a particle and an antiparticle met – when the particle kissed its mirror-image?

After getting a research degree at Glasgow, Touschek went to the Italian laboratory at Frascati, near Rome, where he initially made himself understood using a mixture of Latin and English. In 1960 at Frascati, he had a brain-wave. Under the same electromagnetic conditions, particles and antiparticles move in opposite directions – in 1932, Carl Anderson had been hard put to distinguish between positrons coming down and electrons going up in the magnetic field of his detector. Why not, suggested Touschek, put electrons and positrons into the same accelerator, whirl them round in opposite directions, slightly offset, in the same magnetic field and then crash the two beams together? The colliding electrons and positrons would annihilate each other, producing energetic photons – violent bursts of radiation. The energy of the colliding electrons and positrons could be varied, sweeping across a band of photon frequencies like an ultra-high-energy radio transmitter. If the physicists were lucky, the high-energy photons produced by smashing a high-energy electron and positron together would reach the wavelength needed to make a heavier particle–antiparticle pair, a quark and its antiquark. Any quark–antiquark signal would immediately be 'heard' in the surrounding detector.

To make such an electron–positron collider first meant making positrons. This can be done by thumping a beam of electrons into a block of metal, creating many electron–positron pairs, and then using a magnetic field to sweep all other particles away. After initial proof-of-principle tests at Frascati in 1961, Touschek's AdA – Annello d'Accumulazione, with a diameter of just over a metre, was moved to the French Orsay laboratory near Paris which had a powerful positron source. In 1963, AdA collided a beam of electrons with a beam of positrons. Once AdA had pointed the way, other machines followed. Inching up the energy of the circulating electrons and positrons, physicists were able to supply enough annihilation energy to make other particles – from electrons and positrons came quarks and antiquarks.

Stanford University, California, had initially chosen a different electron route, and a 2-mile electron cannon – the Stanford Linear Accelerator Center, SLAC – became a feature of the landscape near Palo Alto, south of San Francisco. In 1967, this mighty subnuclear 'X-ray machine' revealed the proton's quark skeleton for the first time. However, Stanford also wanted to build an AdA-like electron–positron collider ring. Having constructed one of the world's largest physics machines, SLAC could not ask for too much more funding and built a modest 80-metre diameter electron–positron collider, SPEAR, on a parking lot in the shadow of the big machine. Eighty metres was large compared with AdA, but elsewhere much bigger electron–positron colliders were already on the drawing boards.

Gradually, the energy of SPEAR's electrons and positrons was increased, and in November 1974 SPEAR stumbled on research gold – its annihilation radiation reached a frequency which made the waiting detectors rumble ominously. Carefully fine-tuning the energy, the SPEAR team, led by Burton Richter, homed in on a particle call-sign which almost deafened their detector. The radiation from the colliding

FIGURE 10.1 The SPEAR ring at the Stanford Linear Accelerator Center (SLAC) showed how positrons (the antiparticles of electrons) could be used to make heavier antiquarks (photo SLAC).

electrons and positrons had tuned into a hair-trigger subnuclear state never seen before – a new quark and its antiquark. Heavier than the three quarks Gell-Mann had predicted ten years before, it was a new kind of matter. Continuing the whimsical quark tradition, it was called 'charm'. The same charmed quark–antiquark particle was seen by a group led by Sam Ting at Brookhaven, and in 1976 Richter and Ting were summoned to Stockholm to receive their Nobel prize.

With a fourth quark discovered, how high did the quark ladder go? Reaching for higher and higher energies to ascend this ladder, electron–positron rings got bigger and bigger, and on 14 July 1989, the 200th anniversary of the French revolution, the biggest of them all came into operation, the 27-kilometre circumference LEP (Large Electron Positron) ring at CERN, near Geneva, Switzerland. To precision tune its particle and antiparticle beams, LEP even has to take account of the moon, as land tides bend the Earth's crust and distort the LEP ring ring twice every 24 hours. Even a few centimetres in 27 kilometres is enough to spoil the aim of LEP's electrons and positrons.

COOL ANTIPARTICLES

After Bruno Touschek's electron–positron success, the colourful Russian physicist Gersh Budker had proposed building a ring at the Soviet laboratory at Novosibirsk to hold contra-rotating beams of protons and antiprotons. Gersh Itskovitch Budker, a Jew, preferred to call himself Andrei Mikhailov Budker, as this sounded 'more Russian'. Married five times ('my romances all ended in marriage'), the dynamic Budker had a lifelong commitment to exploit the scientific techniques of the 'wonderful world of particles' for the public good. To make a proton–antiproton collider, Budker was faced with the challenge of building an antiproton source. The trouble with antiprotons, produced at much higher energies than positrons, is that they are much more unruly. As Budker poetically put it, his dream of a proton–antiproton collider was as difficult as 'a rendezvous of the arrows of Robin Hood on Earth and of William Tell on Sirius'.

Proton–antiproton pairs are produced, along with clouds of other particles, when a proton beam is slammed into a target. The problem was that antiprotons, as well as being rare, come out in all directions and with all energies. Even using magnetic lenses and filters to enrich the beam, the anarchic antiprotons jiggle around too much for an accelerator's comfort. Most would fly out of orbit and be lost in the walls of the ring. With antiprotons too 'hot' to handle, Budker realized that, before submitting to the accelerator, nuclear antiparticles would have to be cooled. He proposed leading the unruly antiprotons through a smooth 'sleeve' of electrons, so that the antiprotons' unwanted heat – erratic sideways motion – would be absorbed by the surrounding cold electrons. In 1974, Budker's 'electron cooling' scheme worked.

At CERN, a taciturn Dutch accelerator specialist called Simon van der Meer had another idea for dealing with such anarchic particle beams. He proposed using pickups to monitor a roughly circulating beam. The pickup signal would show how the beam was diverging so that a suitable correction could be calculated. But by this time the beam would have moved on. Van der Meer proposed using feedback electrodes on the opposite side of the circular particle orbit, fed by a wire across the diameter of the ring. For a 10-metre diameter ring, a particle travelling at the speed of light takes 150 nanoseconds (billionths of a second) to go halfway round the ring. To send a signal directly across the diameter takes 100 nanoseconds. If the correction could be calculated inside 50 nanoseconds, the feedback signal would catch up with the portion of beam and nudge the beam into shape. Over many thousands of successive corrections, the beam would gradually be smoothed. In 1972, van der Meer wrote down the idea, called 'stochastic cooling', concluding with the words: 'This work was done in 1968. The idea seemed too far-fetched at the time to justify publication.' Using fast electronics to handle the signal processing, the technique was made to work in 1974.

Keeping track of all this work was a big, ebullient Italian physicist called Carlo Rubbia, who always has an uncanny sense of where physics is heading. In the mid-60s, Rubbia knew that physics understanding had reached a vital crossroads. Physics always looked for unification, to explain as many phenomena as possible with a minimum of underlying theory. Until the early nineteenth century, electricity and magnetism had once been considered distinct. Rubbing ebonite rods seemed to have nothing in common with a compass needle. Slowly it was realized that electricity and magnetism are intrinsically linked – a wire carrying an electric current produces a magnetic field; a wire moved in a magnetic field produces a current. This duality between electricity and magnetism was enshrined in the electromagnetic equations written by James Clerk Maxwell in 1864. Having achieved this satisfying unification, Maxwell's equations also predicted that light was an electromagnetic wave.

A hundred years later, theorists suspected that an analogous unification should link electromagnetism and the force responsible for radioactive beta decay, now called by physicists the 'weak force'. Although the magnetic effect of a current and the beta decay of a nucleus could hardly appear more different at face value, at a fundamental level electromagnetism, light waves and photons had many deep parallels with the weak force. The unification cycle repeated itself when a host of theoreticians, spearheaded by Americans Sheldon Glashow and Steven Weinberg in the US and the Pakistani Abdus Salam, pieced together a larger picture, where electromagnetism and the weak force became different manifestations of a common 'electroweak' effect.

To accomplish this had meant a major rethink of what had become by now one of the central issues of physics, the vacuum. As well as being filled with transient quantum bubbles, the vacuum also had a preferred direction. Using a wedge, a piece of wood will split easily and smoothly along its grain, but it requires much more work to saw across the grain.

Just as electromagnetism is carried by light – photon quanta – so the weak nuclear force too has its carrier particles, although they had to be of two kinds, one electrically neutral, called the Z, and the other electrically charged, the W. The electromagnetic force, working along the grain of the vacuum, operates easily over long distances through its weightless photons. On the other hand, the weak force cuts across the grain of the vacuum and is only felt at very close range. The W and Z carriers of the weak force had to be very heavy, about a hundred times heavier than a proton, heavier even than an iron nucleus.

Like photons, Ws and Zs are at work all the time, but to make them come out in the open means supplying enough energy to create their mass. In the mid 1970s, no electron–positron collider could supply enough energy to manufacture Ws and Zs. Rubbia saw the as yet untested proton–antiproton route as the first real opportunity to get to the Ws and Zs. The protons and antiprotons would not themselves be producing new particles. The protons and antiprotons would be Rubbia's chariots

FIGURE 10.2 Carlo Rubbia, who saw the physics potential of antiprotons (courtesy Photo Boutique/B. Pillet, St. Genis).

of war to bring quarks and antiquarks into the fray. This was more than just a proposal for another experiment. To make antiprotons and bring them into collision with protons would mean adapting a whole big laboratory. Rubbia took his proposal to Fermilab, where a new particle accelerator, the world's biggest, had just started working on the Illinois plain near Chicago. Rubbia's idea was turned down flat.

Shortly afterwards, an experiment at Fermilab discovered the heaviest particle ever seen at the time. This was a heavier cousin of the particle discovered by Richter and Ting in 1974, made of a new quark, the fifth so far, and its antiquark. The tradition for using whimsical names for quarks was now well established, but there was no universal agreement for what the new quark should be called. Some chose the poetic name 'beauty', others the more mundane 'bottom', as an analogy of 'down', one of the two light quarks of which the proton and neutron are made.

After being refused by Fermilab, Rubbia brought his ambitious proposal to CERN, where van der Meer was perfecting his new stochastic cooling scheme. By the mid-1970s, CERN had an impressive array of world-class proton accelerators. The first big European atom smasher, the Proton Synchrotron, or PS, had come into operation there in 1959, and a new machine, the SPS, Super Proton Synchrotron, similar in size to the Fermilab machine, was just coming on line. Building the SPS had been a race with Fermilab, but the US laboratory had won. Providing research facilities for thousands of physicists from all over Europe and even further afield, CERN's heavy committee structure and continental European correctness had always preferred to back 'safe' experiments, leaving the more extrovert Americans to run risks, but also letting them walk off with the Nobel prizes. At first, getting international collaboration to work had been recompense enough, but research has to reap its rewards too. Piqued by this apparent lack of scientific success, CERN management decided to back Rubbia's long shot. It was a bold decision. Too bold, said some. The SPS was only just getting into its stride, and it would be a pity to have to close it for modifications. Proton–antiproton technology was at the limit of technological feasibility. Even if all the new techniques worked, throwing three quarks and three antiquarks at each other at energies never reached before would just produce a mass of confused debris, masking anything new that might be produced.

To test the feasibility of such a scheme, and to weigh the relative merits of Budker's electron cooling and van der Meer's stochastic cooling schemes, CERN built a small test ring called ICE – Initial Cooling Experiment. By 1978, van der Meer's at first sight more difficult stochastic cooling route seemed a better option for taming unruly beams than Budker's. But the ICE tests had been done with protons, and there was one final test to make. Protons live for ever, and so, in principle, should antiprotons. But nobody had ever tested this. Perhaps antiprotons, like the new heavy particles revealed in high-energy experiments, eventually disintegrated. Following the 1965 antiproton discovery, the longest that a single antiproton had been in view was 140 microseconds. There would be no point in building elaborate and expensive antiproton facilities if the antiprotons would promptly disappear. In 1979, the ICE ring received its first taste of antiprotons and the physicists held their breath. The antiparticles were held in the ICE ring for hours. CERN gave the green light to Rubbia's proposal, and colliding protons and antiprotons became a challenge rather than a dream.

A generation of antiparticles had to be created; antiparticles which had to be conceived, born, brought up, schooled, and put to useful work. The heart of the new CERN antiproton project was an antiproton 'factory'. Pulses of raw antiprotons were magnetically selected from the fruits of protons dumped into a target. A million protons produced about one antiproton, as a proton–antiproton pair, but each proton pulse contained ten million million particles. (Later, the raw antiprotons would be focused by a metallic 'lens' powered by thousands of amps of current to concentrate the particles into a narrow beam pipe. Such lenses are made of the light metal lithium to reduce the chances of antiprotons being intercepted as they pass through.)

The antiprotons were fed into a new 'Antiproton Accumulator' ring, there to be tamed with van der Meer's stochastic cooling. Once tamed, each antiproton pulse had to be switched to a second 'stacking' orbit alongside, and another antiproton pulse fed in, cooled and transferred to the stack. After several days and several hundred thousand injected antiproton pulses, a million million antiprotons should be orbiting in the stack, a level comparable to a normal proton beam. At this point, the antiproton stack would be tipped out of the antiproton accumulator and their energy boosted, first in the PS, then in the SPS, all the time in tandem with counterrotating proton beams. When the proton and antiprotons had reached their final 'coasting energy' in the SPS, the beams would be nudged into collision. It was too complicated, said the critics, many from across the Atlantic. It would never work. The delicate antiprotons would be swallowed up.

On 3 July 1980, the antiproton accumulator was first put through its paces in a dry run with protons. With everything working as it should, the currents in the electromagnets switched direction, and the first antiprotons were fed in. But with only a few stochastic cooling units commissioned, initial antiproton intensities were a few per cent of what was eventually hoped for. Meanwhile, new tunnels were being completed to feed the antiprotons from one ring to the next. By the summer of 1981, all the CERN accelerators had been run in for antiprotons and the first high-energy proton–antiproton collisions were recorded. Carlo Rubbia delayed his departure to an international conference in Lisbon so that he would be able to announce that these collisions had been seen. The critics changed their tune. There would never be enough collisions to do any physics experiments. The spotlight shifted from providing antiprotons to recording and analysing the resulting proton–antiproton collisions.

To look for these collisions, Rubbia had initiated the construction of huge detectors, thousands of tons of equipment arranged in concentric boxes surrounding the proton–antiproton collision point, like a huge high-technology Russian doll. Each box was designed to capture one aspect of the collisions, and, by putting the information from all the boxes together, the physicists would arrive at a complete picture. These detectors were installed in huge underground cathedrals around the collision points in the CERN SPS. Operating these huge detectors needed teamwork on a scale never seen before in a scientific experiment, with some 200 physicists involved. Subsequent experiments were to be much bigger, but the proton–antiproton experiments at CERN in the early 1980s set a new scale in scientific collaboration. Responsibility for the different detector components was shared out between the collaborating research institutes and universities. Hundreds of man-years of effort went into the design, assembly and testing of the thousands of units for the detector subassemblies. Wire by wire, and module by module, the complicated electronics was put together, and piece by piece the detectors came together in the participating institutes. As each module passed its acceptance tests, it was shipped to Geneva. The logistics of the operation were immense, and the dimensions of the detectors in some cases were determined by the maximum size of equipment which could be transported to Geneva.

Eagerly the physicists pieced together the pictures of their first proton–antiproton collisions and analysed them. They saw clear signs of the tightly confined sprays of emerging particles – 'jets' – which showed that the quarks and antiquarks deep inside the colliding protons and antiprotons were actually hitting each other. Quark jets had never been seen so clearly. But the physicists knew that it was too soon to look for the long-awaited Ws and Zs. Although the energy was there, the proton–antiproton collision rate was still too low to give a decent chance of Ws and Zs being formed.

In 1982, the collision rates had been coaxed higher, and the curtain was ready to go up on the first real proton–antiproton run with chances of seeing Ws and Zs. But an accident in Rubbia's detector meant that carefully prepared components had to be taken apart and recleaned, and the run had to be postponed. Physicists grumbled, but the delay turned out to be a blessing in disguise. Instead of a series of short runs interspersed with other physics, the collider programme was shoe-horned into one long run later in the year. It took time to get an antiproton act together, and continuity was not improved by chopping up the show. With one long run, the accelerator specialists were able to perfect their act and ensure that their antiproton supply was more reliable.

In the summer of 1982, British Prime Minister Margaret Thatcher, on holiday in Switzerland, paid a private visit to CERN. With her background as a research chemist, she had always taken a strong interest in basic research. During her visit, she was told 'with a bit of luck, some help from our accelerator colleagues, and a firm belief in Santa Claus, we may have the W for Christmas'. After the visit, she asked CERN's Director General at the time, the German physicist Herwig Schopper, to let her know immediately the Ws and Zs were found. She did not want to have to rely on press reports, she said.

A few months after Margaret Thatcher's visit, the proton–antiproton collision score had been boosted a hundredfold compared with the previous year. The circulating antiproton beams became more and more durable, one antiproton shot being held captive for almost two days before it finally frayed apart. By this time the experiments had seen several thousand million proton–antiproton collisions. Knowing exactly what they were looking for, the experimenters had prepared digital

FIGURE 10.3 August 1982 – UK Prime Minister Margaret Thatcher learns about antiproton developments from CERN Director General Herwig Schopper (right) and Carlo Rubbia's deputy Alan Astbury (left) (photo CERN).

expressways to pile the data through their computers. Preprogrammed electronic turnstiles were set to click each time a collision with a W entry ticket arrived. Although nobody said much at first, slowly some satisfied smiles were being seen around the experiments' control rooms. In December 1982, Schopper was confident enough to tell Margaret Thatcher. Unaware that CERN's supremo had prematurely leaked the news to Downing Street, the British physicist Alan Astbury, Carlo Rubbia's deputy, and Rubbia cabled Thatcher in January 'we begin to reveal the existence of the long awaited W, and incidentally have confirmed beyond doubt the existence of Santa Claus'. The characteristic reply is shown below.

10 DOWNING STREET

THE PRIME MINISTER

26 January 1983

Dear Dr. Astbury

Thank you for the telex which you and Carlo Rubbia sent on behalf of the UK contingent in the UA1 experiment. I'm not sure which is more exciting: the glimpses you have had of the W particle, or the knowledge that Santa Claus really does exist. Anyway, my warm congratulations on a very important discovery, and I am delighted that British scientists were once more in the winning team. I am sure the prize will be confirmed by your experiments in the spring, and that this will be just the first of many important discoveries for your team.

Yours sincerely

Margaret Thatcher

FIGURE 10.4 Letter from UK Prime Minister Margaret Thatcher to physicist Alan Astbury after he had leaked the news that proton–antiproton experimenters had made a major discovery.

Early in the new year, ten candidate Ws could be held up for inspection, but, with the implications of the discovery so momentous, no claim was yet made. But, over the weekend of 22–23 January 1983, Carlo Rubbia became more and more convinced – 'they look like Ws, they feel like Ws, they must be Ws', he remarked. The announcement that the electrically charged W had been seen was made on 25 January.

The experimenters knew that the Z, the electrically neutral companion of the W, needed more proton–antiproton collisions than the W, but, on the other hand, its 'fingerprint' should be easier to spot. In April 1983, a new antiproton run began. Although more antiprotons were being formed, collected and accelerated, the Z did not seem keen to show itself, but, as May went by, the first Z calling cards were seen. On 1 June, the Z announcement was made. Out of quarks and antiquarks had come the carriers of the weak force. Physicists called them 'intermediate bosons'. With more imagination, press called it the discovery of 'heavy light', alluding to the deep connection between the new particles and electromagnetic radiation. For the discovery, Carlo Rubbia and Simon van der Meer were awarded the Nobel Prize for Physics in 1984, a short wait by Nobel physics standards.

Going for the W and Z had been a scientific and technological gamble. Sheldon Glashow, Abdus Salam and Steven Weinberg, the three theorists who had pieced together the electroweak theory, had been awarded the Nobel prize in 1979, a courageous move by the Royal Swedish Academy of Sciences in view of the fact that the key W and Z particles predicted by the theory had not yet been seen. As one senior physicist hearing the news commented: 'Does this mean they will have to give back the prize if the W and Z are not found?' The question remained a hypothetical one. Physicists all over the world applauded the achievement. A Fermilab physicist described the discovery as the 'particle physics equivalent of the 1969 Apollo moonshot'.

Its physics mission complete, CERN's proton–antiproton collider was decommissioned in 1992, by which time it had handled several hundred billion antiprotons. Producing these antiparticles had been a major effort in money, time and manpower, but, if that many antiprotons could be brought together at once, they could all be seated comfortably on a pinhead, and weigh less than a speck of dust. With its mission at CERN complete, equipment from the antiproton source was shipped to Japan for a new lease of life generating antiparticles in the Far East.

Fermilab, which had turned down Rubbia's initial antiproton proposal, later decided to follow CERN's suit. The Fermilab proton–antiproton collider started operating in 1985, and had the advantage of beam energies several times higher than those of CERN. The Fermilab collider's big moment came in March 1995, when experiments discovered the sixth quark, by far the heaviest of all, called 'top'. Out of colliding light quarks and antiquarks had come heavy quarks and antiquarks.

In the laboratory, particle–antiparticle collisions, whether of electrons and positrons or of protons and antiprotons, produce more particles and antiparticles, but so far the number of produced particles is always the same as the number of antiparticles. Energy is a measure of temperature, how fast the component particles of matter move, and the energy of these colliding particle–antiparticle beams re-creates temperatures not seen since the first fraction of a second after the Big Bang that created the Universe. Under the laboratory conditions explored so far, the Universe would have contained as many particles as antiparticles. But we see only particles around us. Particle–antiparticle collision experiments are striving to attain higher temperatures in the search for the delicate asymmetries that break the antiparticle mirror described in chapter 8. If so, they will be simulating the Big Bang conditions which destined that matter would reign over antimatter.

11 Setting a trap for antimatter

Antimatter annihilates with matter, so any attempt to store anti-matter in a physical container is doomed. Antimatter can only be kept in a box without material walls. One way of doing this is to magnetically confine circulating beams of positrons or antiprotons, as described in the previous chapter. But this can only work if the antiparticles are flying round at high speed, the centrifugal force making them want to fly off at a tangent being exactly balanced by an inward magnetic pull. What about antiparticles moving too slowly to go into such magnetic orbits? Can these be stored?

In 1984, a team led by Hans Dehmelt at the University of Washington, Seattle, succeeded in holding a single positron captive for three months in a specially designed particle 'trap' – a tiny cylinder a fraction of a hair's breadth in diameter and length in which the lone particle quietly rested on cushions of electric and magnetic fields. But Dehmelt had not initially designed this device as a trap for antimatter. There was a much more fundamental physics reason why he had toiled for almost twenty years to build a particle container without walls.

When a string vibrates, it imparts energy to the surrounding air, and this energy is picked up by our ears. Electromagnetic waves, such as light, are vibrations which radiate energy of a different kind. To make vibrations first needs some kind of oscillator, like a string fixed at both ends, and then the oscillator has to be made to 'twang'. If electromagnetic energy were produced by oscillations of tiny invisible strings, then in principle there would be no limit to the frequencies of the oscillations. For sound, higher frequency means higher pitch, while, for electromagnetic radiation such as light, higher frequency means moving away from the red end of the spectrum and towards the violet.

In 1900, Max Planck realized that atomic oscillators cannot give out energy continuously (see chapter 6) – atomic oscillators cannot become infinitely small, and ultimately the high frequencies come up against a barrier. A stringed instrument like a violin can be plucked, giving a short, isolated note, or bowed, giving a continuous note. But, in slow motion, even bowing is not continuous – the bow rubs across the string in a continuous series of tiny plucks which simulate a continuous note. Planck said that atomic oscillators, too, have to be plucked, each pluck releasing a quantum of radiation energy. With enough oscillators, the resulting vibration appears continuous, in much the same way that we cannot detect to-and-fro bowing from the sound of the assembled violins of an orchestra. Moreover, said Planck, the energy needed to 'twang' the atomic oscillators depended on the frequency emitted – the higher the frequency of the emitted electromagnetic 'note', the more energy was needed to work the oscillator. Planck wrote down the simplest possible equation to ensure this: $E = h\nu$, where E is the oscillator energy and ν is the frequency, the two being related by the number h, Planck's constant.

The frequency of a vibration is the number of wave crests passing per second. The distance between successive wave crests is the 'wavelength'. For vibrations moving at fixed speed, like sound or light, the longer the wavelength, the fewer wave crests pass per second – the lower the frequency. The smaller the wavelength – the higher the frequency. The bass strings of a harp are longer than those of the high notes. In Planck's language, the energy needed to pluck the strings of a quantum harp also changes – in such a harp, the low-frequency bass notes can be softly tapped, the high ones have to be thumped with a hammer.

A quarter of a century later, this deep connection between energy and radiation worked its way into physicists' consciousness. Werner Heisenberg, the inventor of matrix mechanics, stopped for a while to ponder over the implications of the bizarre new quantum results predicted by his matrix manipulations. Why did Nature appear to be so quirky on the subatomic scale? With experiments on individual subatomic particles then impossible, Heisenberg imagined a subatomic situation – a hypothetical 'thought experiment'. The problem was to determine simultaneously the position and speed of a single subatomic particle – seemingly a reasonable enough request.

The first thing to do would be to look at the particle. To do so means shining a light on it, and the reflected light would show where the particle is. But the electromagnetic vibrations of this illumination have a certain wavelength, and to a certain extent this wavelength masks the accuracy of the particle position measurement. The wavelength of ordinary light is less than a millionth of a metre. An atom is about ten thousand times smaller than this wavelength, so that a single atom is much too small to be seen by light – the huge light waves just surge past and over the atomic obstacle. To see an atom needs much smaller wavelengths. How can these be obtained?

Planck's quantum picture showed that radiated energy is not continuous, although, with enough atomic oscillators at work, the net result is as though it were. Radiated energy is like rainfall – everything gets wet even though the water falls as individual raindrops. Planck showed that low-energy (low frequency – long wavelength) radiation is a fine mist which makes the ground uniformly moist over a wide area, while high-energy (high frequency, short wavelength) radiation resembles more the hail of a thunderstorm, which can wreak enormous damage.

But, if quantum radiation is like raindrops or particles, then particles could also look like radiation, proposed Louis de Broglie boldly in his 1923 designer equation. The higher the energy of the particles, the higher the frequency and the shorter the associated wavelength, said de Broglie. A beam of electrons has a wavelength that depends on its energy, and this was eventually exploited in electron microscopes to reveal viruses too small to be seen by ordinary light, and details of molecular and atomic structure.

Suppose, said Heisenberg, we want to build such a microscope capable of pin-pointing a single electron. Such a refined position measurement has to use as small a wavelength (as high a frequency) as possible, so the energy of the hypothetical electron microscope has to be pushed up and up until the subject electron suddenly comes into focus. We measure where the electron is. If we can also measure the electron's velocity, in principle we should know where the electron will be at any time and be able to predict its future. But things are not so simple. The simple act of seeing the electron means that a quantum of electron microscope radiation is reflected from the object electron, bouncing back into the 'eyepiece'. Radiation capable of seeing an object as small as an electron has a lot of energy. When this radiation bounces back from the object electron, the bounce is so hard that the object electron recoils sharply away, changing its velocity. We may have measured exactly where the object electron once was, but in doing so have changed its fate.

Likewise, if we wanted to measure the electron's speed, for example by timing how long it takes for it to cross a certain reference gap, we cannot know exactly where the electron is any more. The ultimate case is when the electron is at rest. If it has no speed, we cannot time how long it takes to cross our reference gap, so the electron can be anywhere! This conundrum is known as the Heisenberg Uncertainty Principle. It took fifty years before technology caught up with Heisenberg's 'thought experiment' and physicists were able to approach these conditions for real.

'LESS IS MORE'

Brought up in Berlin, Hans Dehmelt was drafted in the Second World War and narrowly escaped from Stalingrad, but was later taken prisoner in 1945. After being released in 1946, Dehmelt made a precarious living fixing old radios before entering the University of Göttingen to study physics. At Göttingen in 1947, Dehmelt was one of the pallbearers at the funeral of the stalwart Max Planck whose eldest son was killed in the First World War and whose two daughters died when they were young. During the Second World War, Planck's house was bombed and he lost most of his personal papers, while his remaining son was executed for his part in the attempted assassination of Hitler in 1944. After the war, the durable Planck, then aged 87, briefly became president of the former Kaiser Wilhelm research institute, renamed the Max Planck Institute in his honour.

As a child, Dehmelt had heard his father expound on the intricacies, beauty and wisdom of Roman law. But laws made by humans were too abitrary for Dehmelt. 'I felt myself under the spell of the wonders of the physical world and its universal and immutable laws', he said later, a desire which led him in 1978 to become one of the first humans to see a single atom. Dehmelt's dedicated development of containers without walls brought the technology of handling electrically charged particles to a fine art, and eventually brought him an invitation to Stockholm in 1989 to share the Nobel Prize for Physics. These containers also opened up the possibility of suspending particles of antimatter in space, preserving them from the dangers of annihilation with the surrounding matter.

'The phenomena of the physical world are extraordinarily rich and interwoven', says Dehmelt. 'Thus the life of an experimenter is spent attempting in the laboratory to artificially create phenomena simple enough that they can be analysed.' As a physics student in Göttingen, Dehmelt remembers a lecturer putting a chalk dot on the blackboard and saying 'Here is an electron.' This impressed Dehmelt, who had also heard lectures by Werner Heisenberg at Göttingen in which the master of quantum mechanics had explained that, according to his Uncertainty Principle, an electron at rest, with zero speed, could be anywhere! From his knowledge of electronics, Dehmelt knew that electrons can be controlled by electric and magnetic fields, and set himself the challenging goal of trapping subatomic particles for long enough to beat the Uncertainty Principle at its own game.

When a high voltage is applied to a tube of gas at low pressure, the gas atoms break apart and a continuous column of electrons – 'cathode rays' – marches relentlessly away from the cathode. This march of the charged particles can be affected by a magnet placed next to the tube. The early experiments on cathode rays had shown how the glow of the cathode rays could be bent by a magnet. Suitable arrangements of magnetic fields could make electrons execute more complicated gymnastics. Lawrence's cyclotron could hold electrons captive in a fixed orbit, the inward magnetic grip exactly balancing the centrifugal skid of the

FIGURE 11.1 Hans Dehmelt – particle trapper (photo CERN).

rotating particles. In a cathode ray tube, if the magnetic field is made strong enough, the electrons are bent right round, and the usual current cut off . These electrons, not knowing where to go any more, meander around, bumping into residual gas molecules in the tube and losing energy. These lower-energy electrons are no longer balanced by the magnetic force and loop back towards the anode. In the 1930s, F.M. Penning showed how this principle could be used to measure how many residual gas molecules there were, and invented a very useful vacuum gauge. These Penning gauges look very much like the large valves once used in radios.

In 1956, Dehmelt took such a Penning gauge and changed its voltage so that electrons, instead of eventually reaching the anode and registering a current, meandered around permanently in the magnetic field. In such a tube, the electrons roll around in the electric and magnetic fields like marbles in a rotating bowl. The arrangement of electrodes resembles a tin can cut open – the cylindrical part being the anode and the separated top and bottom of the can as twin cathodes. If an electron strays near the cathode 'lids', it is pushed back into the can, becoming trapped in a lacework-like orbit. This device Dehmelt called a Penning trap, acknowledging the electronics pioneer without whom the idea would not have been possible. If Dehmelt had not chosen to acknowledge Penning, the device would surely have become known as a Dehmelt trap. The electrons rotating in the trap also act as tiny radio transmitters, and Dehmelt was able to pick up this radiation in a surplus Navy radio receiver. As the electrons radiate, they lose energy, rolling less vigorously in the 'bowl' of electric and magnetic fields.

Like many physicists, Dehmelt was fascinated by the almost mystical power of the Dirac equation. It was a 'A thing of power and beauty', he said. According to this equation, even an electron at rest has angular momentum, as though it were spinning around its own axis. Moreover, this axis cannot point anywhere; like some kind of quantum switch it can only point meaningfully in two directions – up or down. Dehmelt could not understand how an electron, with no visible dimensions, a mathematical point, could rotate at all. He was not the first to have trouble with the idea of something with zero dimensions still managing to rotate – Pauli described electron spin as the 'classically nondescribable two-valuedness' of the electron. The electron does not spin at all, said Pauli, it just has some property that simulates rotation.

But, in the absence of any other picture, the idea of spin as physical rotation persisted. The rotation, said physicists, was somehow intrinsic. If a vessel containing liquid is made to spin at high speed, the liquid continues to spin when the vessel is suddenly clamped. But all these pictures required the electron to have inner dimensions, whatever they were. When an electrical charge rotates, it behaves like a magnet. Likewise the electron, which rotates in some way, should also act as a tiny magnet. Early experiments had shown that the spinning electron has twice the expected magnetism, underlining the suspicion that the electron spin was somehow different. One of the triumphs of the Dirac equation was its explanation of the electron's double magnetism, a direct result of relativity. Later experiments showed that the ratio was not exactly two – there was a tiny difference of 0.1 per cent, attributable to the same sort of quantum effects that produced the Lamb shift (chapter 6). The electron is surrounded by a fuzzy cloud of attendant photons and electron–positron bubbles, and these leave their mark on the electron's magnetism. The Dirac equation cannot answer for these effects, which call for the quantum electrodynamics of Richard Feynman and Julian Schwinger.

To measure the magnetic effects of atoms, in 1920 Otto Stern had devised an ingenious method of shining a beam between the poles of a specially shaped powerful magnet. In this magnetic field, the atomic magnets line up like tiny compasses. However, these quantum compasses can only point in certain directions, in much the same way that atomic electrons can only sit in special orbits, the rungs of an energy ladder. As each particle passed through Stern's magnet, it switched into one or other of the quantum-allowed magnetic directions. In this way Stern measured the magnetism of the proton, showing it to be very different to that of the electron. This earned Stern the Nobel Prize for Physics in 1943, the first time it had been awarded since the outbreak of the Second World War.

However, Stern's technique did not work for electrons. The much lighter electrons curve in the magnetic field and the magnetic effects of this rotation mask any attempt to measure the electron's intrinsic magnetism. To overcome this, Dehmelt held an electron in a trap cooled by liquid helium to freeze out thermal jiggles and then applied a tiny magnetic field. The resulting radio transmissions, concentrated around the single note of the circulating electron, showed a steady background noise due to residual thermal motion, but from time to time the fundamental note changed frequency as the electron flipped its spin direction from pointing up to down and vice versa.

As well as Dehmelt's electromagnetic Penning trap, there was another possibility, using only electric fields. Instead of capturing the electrons in a magnetic orbit, the idea is to use a high-frequency oscillating electric field like a tuning fork. Those particles naturally in tune with the high-frequency 'fork' will resonate – vibrate in sympathy – others will not. The spectrum of this induced radiation shows what atoms are present. This technique, invented by Wolfgang Paul in Bonn in 1954, became useful for atomic analysis. Paul had run a laboratory class at Göttingen when Dehmelt had been a student, and Paul accompanied Dehmelt on the Stockholm prize rostrum in 1989 for their complementary work on particle traps.

To improve his technique and make precision measurements rather than just monitoring the steady flip-flop of electron spin, the ingenious Dehmelt shone a bright laser beam into a trap. Just as the note of a police-car siren appears to change as it goes past, the 'Doppler effect', so the circulating particle sees a band of laser frequencies, stretching from a maximum when the particle is travelling towards the beam, to a minimum when it is travelling away from the beam. The width of this frequency band depends on the orbital speed of the particle. By tuning the laser, this band can be made to coincide with any internal rotational frequency of the particle, which then resonates in sympathy and gives out radiation. In this way, Dehmelt was able to make precise measurements of the electron's magnetism, showing that it is 2.0023193044 times what was originally expected, a 0.1 per cent difference from the factor of two predicted by the Dirac equation. Doing quantum electrodynamics calculations à la Feynman predicted 2.0023193048! These are among the most striking agreements between theory and experiment ever made, the few parts in a billion accuracy being equivalent to a sniper on Earth hitting a coffee cup on the Moon.

A single electron 'stuck' to the Earth by electric fields can be likened to a giant atom, with the Earth as its nucleus. Dehmelt called this pseudo-atom 'geonium'. After a decade of valiant effort, Dehmelt and his team precisely measured the magnetism of both the electron and, by reversing electric fields, the positron. Never had so much effort been put into investigating such a small experimental system. Dehmelt called the project 'Less is more'. The measured magnetism of the electron and its antiparticle is the same to a few parts in ten million million, the most accurate check so far that charged particles and antiparticles behave the same way.

This magnetism is 1.000000000055 times that predicted by the best physics calculations. This tiny difference between the predicted and measured values suggested to Dehmelt that the electron and positron might not be infinitesimally small point particles, but instead have a finite size. Using the measured magnetism as a benchmark, the electron would be a hundredth of a billionth of a billionth of a centimetre (10^{-20} cm) across, a thousand times smaller than the finest 'X-ray' of an electron yet made. With this factor of a thousand as an insurance policy, Dehmelt speculated that at this scale the electron would no longer be a single particle, but would be made up of subquarks, each ten billion times heavier than an electron, another rung on a descending ladder which ultimately terminates at what he calls the 'cosmon', the heaviest, and smallest, particle the Universe has ever seen.

TRAPPING ANTIPROTONS

With electrons and positrons having been trapped and measured with unprecedented precision, attention then turned to antiprotons. Antiprotons had been confined in CERN's high-energy machines, where the idea was to exploit them rather than measure them. CERN also built a low-energy antiproton ring, LEAR, where the objective was to study the antiproton as well as to use it. (LEAR was a breakthrough in antimatter research and is the subject of the next chapter.) Gerald Gabrielse, once a colleague of Dehmelt at the University of Washington, proposed using a Penning trap to collect LEAR's antiprotons. Many problems had to be overcome *en route*. To minimize the thermal jiggling which can upset the delicate trapping orbits, the particles have to be cooled to the equivalent of liquid helium temperatures and even lower. But liquid helium would quickly eat up the precious antiprotons, so a gas of cold electrons has to be used instead, the negatively charged electrons and the negatively charged antiprotons being able to coexist peacefully. Antiprotons from LEAR were lured into a 13-centimetre-long ultra-high vacuum Penning trap. Once inside, the voltage at the entrance electrode is altered, shutting the trap 'door' and preventing any more antiprotons from entering. Those already inside were sealed into a tiny space of a cubic millimetre.

At this stage, the trap contained about ten thousand antiprotons together with a much larger number of electrons from the cooling gas. Briefly pulsing the trap's electric field ejected the lighter electrons. By gradually reducing the depth of the trap's electromagnetic 'bowl', excess antiprotons gradually spilled out, until just 15 remained. At this stage, their different rotations enabled them to be electromagnetically weeded out one by one, until just a single antiproton remained. The trap depth was then increased to make sure the lone particle of anti-matter did not escape. Tuning into its tiny radio signal, Gabrielse's team measured its frequency. The first antiproton was trapped by Gabrielse in 1986. Having made these measurements, attention turned to accumulating ultra-cold antiprotons in a deep trap, and by 1993 Gabrielse's team had been able to store about a million at a time.

To develop these particle traps, physicists had to tax their imagination and ingenuity. But Nature is even more ingenious and provides its own way of trapping particles. In atoms, particles of opposite negative charge – protons and electrons – are locked together. In principle, atom-like systems can be made using ordinary nuclei and any negatively charged particles, such as antiprotons. To make antiprotons, a high-energy beam first has to be smashed into a target and the secondary particles, including any antiprotons, emerge at very high speeds, a good fraction of the speed of light. These fast-moving antiprotons crash into ordinary atoms, slowing down as they knock out electrons. The antiproton eventually slows to a 'walk'. When such a slow antiproton encounters a lone proton, the particle and its antiparticle can become locked in orbit round each other by their electromagnetic attraction. This atom, called protonium, is a hydrogen atom in which the orbital electron is replaced by an antiproton. Protonium – Nature's own way of trapping antiprotons – was first seen in experiments at CERN in 1970. However, because the antiproton is some 2,000 times heavier than the electron, the protonium atom is much smaller than that of hydrogen. Protonium, like all other atoms, has a ladder of allowed energy levels, and initially the antiproton alights on one of the higher rungs of the energy ladder. As it continues to lose energy by radiation, it gradually steps down the energy ladder and approaches the proton. In hydrogen and other everyday atoms, the nucleus is so much heavier than the orbital electrons that it can be considered fixed, the centre of the atomic 'solar system', with the distant electrons in orbit around it. However, in protonium, the two captive particles have the same mass. One particle cannot be considered the nucleus and the other to be in orbit around it. Instead, they circle warily round each other. When the antiproton attains the lowest rung of the energy ladder, the orbit of the proton and antiproton is comparable with the size of the particles themselves – the proton and the antiproton actually overlap with each other and the two particles soon annihilate. But the proton and antiproton still manage to circle each other briefly in an electromagnetic *pas de deux* before their annihilation swan-song.

Until 1987, physicists had never seen an antiproton attain the lowest possible rung of the protononium energy ladder, as the antiparticle had always succumbed to the proton *en route*. In 1987, physicists working at LEAR followed the cascades of X-rays released as an antiproton ventured closer and closer to a proton. In one chain in a hundred, the antiproton actually made it to the bottom of the energy ladder. The properties of the protonium atom can be accurately calculated, but these predictions are spoiled when the proton and antiproton begin to overlap. The resulting shifts in atomic properties enable physicists to see what happens when a proton and an antiproton briefly co-exist before succumbing to their fate of annihilation.

With various kinds of traps at their disposal, artificial or natural, by 1986 physicists had isolated single antiprotons and single positrons. In principle, they had the ingredients for making the first atoms of true antimatter. Electrically neutral atoms consist of negatively charged electrons orbiting round a positively charged nucleus containing protons and neutrons. For antimatter, the electrical roles would be reversed – positively charged positrons would have to orbit around negatively charged antinuclei consisting of antiprotons and antineutrons. Hydrogen has the simplest atom of all – a lone electron orbiting a single nuclear proton. The simplest antimatter atom – antihydrogen – would have a positron orbiting a nuclear antiproton. Drawing on the experience with single particles, the route to making antihydrogen seemed to be mapped out – introduce antiprotons and positrons into the same electromagnetic trap, and let them snare each other.

The resulting antimatter atoms, electrically neutral, would immediately fall out of the trap under gravity. But, when an antiparticle feels gravity, does it fall down at the same rate as ordinary matter? According to Einstein, gravitational pull depends only on mass, so that antimatter should fall down in the same way as matter. But nobody has ever seen how antimatter behaves under gravity. Perhaps it could even 'fall up'! Watching the effect of gravity on antimatter would be the next step in a series of experiments which Galileo had begun four centuries before.

To investigate how antimatter behaves under gravity, antiproton physicists are building large Penning traps to capture millions of antiprotons and measure the effect of the Earth's gravity on them. This will be a very delicate test of antimatter and gravity. Perhaps antimatter and matter are pushed apart by gravity. (Another crucial test would be to study the way antiprotons fell to an Anti-Earth. Powerful theorems say that an antiproton falls to an Anti-Earth in exactly the same way that a proton falls to Earth, but simulating an Anti-Earth would be much more difficult than making antiprotons!)

We are conditioned to the idea of gravity being a universal attraction – every mass in the Universe attracts all the other masses. But this has not always been the case. For the Universe to have reached its present size, the gravity that emerged after the Big Bang would have to have been a repulsion, enormously more powerful than the attraction we now know. Could antimatter have played a role in this 'antigravity'?

12 Glue versus antichemistry

The SPEAR ring at Stanford showed how profitable it could be to hurl electrons and their antiparticles at each other and create quarks and antiquarks instead. The electron–positron route looked a good research bet. One of the first to follow SPEAR was the Deutsches Elektronen Synchrotron – DESY – laboratory in Hamburg. DESY already had electrons on tap and started a family of electron–positron rings.

The first, DORIS (DOppel-RIng-Speicher, or double ring storage), a 50 × 100 m oval, came into action just too late to catch SPEAR's fourth 'charm' quark. Hurling its electrons and positrons harder at each other, DORIS was able to ascend the next rung of the quark ladder. DORIS spawned particles made of the fifth 'beauty' quark and its antiquark.

Even while DORIS was still getting into her stride, DORIS' daughter, PETRA, was conceived. She was 730 metres in diameter and with higher electron and positron energies, so perhaps PETRA could be first to reach the sixth rung of the quark ladder. PETRA's particles and antiparticles began colliding in 1978, nine months earlier than scheduled, and a full two years ahead of the US opposition. PETRA had the elecron-positron field to herself.

The rungs of the quark ladder are not evenly spaced and nobody knew how high the next step was. PETRA's beams carefully reached out, probing for the first signs of a new quark–antiquark foothold. But this was a long shot. The energies of PETRA's electrons and positrons would not be able to manufacture the W and Z weak interaction carriers. That big prize had to await Carlo Rubbia's heavy proton–antiproton artillery. What research fruit could be within PETRA's easy reach?

THE QUARK ADHESIVE

When electrons and positrons tune in to a quark–antiquark resonance, the quark–antiquark signal rings loud and clear, as it had done for SPEAR in 1974. However, when the electrons and positrons have energy to spare, the quark–antiquark call sign is obscured. Extra energy is taken up by the quantum bonds which bind the quark and antiquark together. As the energy is increased, these bonds begin to vibrate, like a thick piece of elastic when suddenly stretched. Keep supplying energy and the elastic eventually snaps, the two halves flying off in opposite directions.

Even though a quark–antiquark bond has been broken, this does not separate the quark and its antiquark. A horizontal rod has two ends, left and right. Sawing the rod in half gives two smaller rods, each with a left and right end. Left and right are attributes, not objects in their own right. No amount of cutting can produce a rod with only a right-hand end. In our world, quarks and antiquarks are also attributes – nobody has yet seen a free quark or antiquark. Severing the elastic bond between a quark and its antiquark gives two quark–antiquark pairs, each with its own bond. Initially, the debris coming from electron–positron annihilations clearly displayed diametrically opposite sprays of particles (Figure 12.1), the remnants of ruptured quark–antiquark bonds.

The elastic holding the quark and antiquark together is made up of particles called 'gluons', so called by Murray Gell-Mann because they make quarks stick together. When the quark and antiquark huddle close together, this gluon bond is almost undetectable. Like a long spring, it is inert when its ends are close together, but becomes stiffer as the ends are separated. The greater the extension, the stiffer the elastic spring becomes.

When it splits into two portions with quark and antiquark ends, the overstressed elastic could also radiate some of its stored energy, like the sudden 'ping' of a guitar string as it breaks. Unlike a broken guitar string which releases its tension as sound energy, the tension released by the broken quark–antiquark bond produces particles, gluons. The

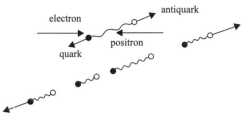

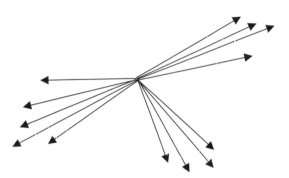

FIGURE 12.1 When a quark and an antiquark are created by the annihilation of an electron and a positron, a quark–antiquark link can snap, producing more quark–antiquark pairs. These are seen as two back-to-back sprays or 'jets' of particles.

FIGURE 12.2 Three clearly defined 'jets' of particles produced by the PETRA electron–positron collider showed that the electrons and positrons could produce gluons, the material that makes quarks and antiquarks stick together. When a quark–antiquark link snaps, as well as producing more quark–antiquark pairs, the released energy can also sometimes spit out an isolated gluon. The gluon then goes on to produce quark–antiquark pairs.

three emerging patches of particles – two portions of gluon elastic with a quark and an antiquark on each end, plus a bunch of gluons – would share the available energy, giving a pattern resembling the three-pointed star of the Mercedes trade mark (Figure 12.2). In 1979, experiments at PETRA saw the first examples of such patterns emerging from electron–positron energy. PETRA could not see the sixth quark, but out of her beams of matter and antimatter had come the glue that held quarks and antiquarks together.

ANTIPROTON SIDESHOW

Gluons make quarks stick to antiquarks, but they also make three quarks stick together to form protons, and three antiquarks to form antiprotons. Exploring the gluon needed low-energy protons and antiprotons. As CERN geared up for its major attack on the W and Z particles by slamming together high-energy protons and antiprotons, could some of these antiprotons be spared for other uses? By putting a particle accelerator into reverse gear, these antiprotons could be slowed down to a walking pace. Combining these low-energy antiprotons with protons, could gluons be coaxed into showing themselves in other ways?

CERN was initially reluctant to spare its precious antiparticles for what seemed like a physics sideshow, but a low-energy antiproton ring, LEAR, was eventually authorized on condition that it had to live on a subsistence antiproton diet. LEAR's kit of parts included four 8-metre-long conduits arranged as the sides of a square, joined by four antiproton channels which magnetically bent the beam through 90°. LEAR had its first taste of antiprotons in 1982. Investigating quark glue was high on LEAR's initial research agenda, but LEAR had to wait until almost the end of its career before making the headlines, and not through studying quark glue.

With a ration of only 6 per cent of CERN's antiproton yield, LEAR had a deprived childhood. This meagre diet meant that LEAR had to learn how to make the most of its precious antiparticles, and the machine specialists developed a method in which just a single antiproton at a time could be removed from those circulating in the ring. With the antiprotons moving almost at the speed of light, shaving off antiprotons this way for 10 minutes gave a beam that resembled a slender antiproton chain stretching from the Earth to the Sun with a single antiparticle every 100 metres.

LEAR had been designed to be used as a low-energy proton–antiproton collider, with counter-rotating beams of particles and antiparticles, but this option was never used. It would have disturbed the clean antiproton beams needed for LEAR's external experiments, custom-built detectors nibbling at the peelings from the edge of LEAR's stored antiproton beam. To probe the annihilation process without interfering with the peripheral experiments, LEAR resorted to a powerful jet of gas, a thick curtain of hydrogen, mainly protons, squirted across the path of the circulating antiprotons. Such a jet is millions of times denser than any circulating proton beam, and provides correspondingly higher levels of proton–antiproton collisions. Although the gas curtain interferes with the smooth circulation of the antiprotons, it only does so at one place and the resulting wobbles in the beam can be smoothed. These gas jets demand sophisticated technology, with ultra-clean hydrogen squirted at high pressure through a micronozzle a micron across. To maintain the high vacuum in the ring, the fine jet of gas, which expands to occupy about a square centimetre, has to be sucked out again on the far side. The squirts are so powerful that the jet emerges from the nozzle at supersonic speeds, at a rate of some 100,000 litres per second. Using the gas jet target was the 'Jetset' detector, a compact 2-metre-high cylinder of high technology. Jetset came into action in 1990, when LEAR had already been in operation for eight years.

By this time, CERN no longer had a world monopoly on antiprotons. Across the Atlantic, at Fermilab near Chicago, where Carlo Rubbia had first proposed his antiproton scheme in 1976, antiprotons had been added to the physics menu. After having slammed the door in Rubbia's face, Fermilab had a rethink, and in 1979 began to push for its own antiproton scheme. Knowing that CERN had a head start, Fermilab, which had the advantage of higher energy beams, chose a longer-term option. Fermilab's trump card was a new machine, the Tevatron, which could take beams to twice the energy of CERN. Fermilab's Tevatron proton–antiproton collider started operating in 1985. Its big moment came in March 1995, when it discovered the sixth quark, by far the heaviest of all, called 'top'. About three hundred times heavier than the fifth quark, no wonder it was never seen by PETRA's electrons and positrons.

Although Fermilab had no equivalent of LEAR, low-energy precision antiproton experiments could be set up around the accumulator ring where the precious particles of nuclear antimatter were stored before being passed to the Tevatron. One of these experiments too used a gas jet target. Fermilab experiments are allocated a sequential identification number. The gas jet target was E760.

In 1991, CERN's SPS proton–antiproton collider was closed for good, and the baton for exploring high-energy antiprotons passed to Fermilab. But CERN still had its antiproton factory, and now its only client was LEAR. Living in the shadow of the big SPS proton–antiproton collider, LEAR had developed an inferiority complex. LEAR's physics trophy cupboard was embarrassingly bare. With research money tight, there were growing fears that CERN's antiproton source would be closed for good and components given to Japan as a goodwill present. This could be prevented if LEAR pulled off a major research success.

One LEAR speciality was synthesizing atoms with antiprotons. The simplest antiproton atom of all is antihydrogen, with a nuclear antiproton and an orbital positron. If positrons could be introduced into the LEAR ring alongside the circulating antiprotons, then antihydrogen could be produced. Chemistry would enter the mirror world of antimatter. LEAR's designers had even foreseen this possibility. The negatively charged antiprotons are steered round the corners of the LEAR square by magnetic fields and groomed by cooling devices in the straight sides. If any neutral particles, like anti-atoms, were formed in these straights, the particles would no longer feel the grip of the bending magnets and would fly out of the ends of the straight sections. The LEAR design included windows in the 90° bends so that any such escaping neutral particles could be picked up by detectors waiting outside the ring.

Despite this forethought, the holes through which anti-atoms could escape were quickly blocked by other instrumentation. Although antiparticles had become everyday physics, most physicists were not interested in antichemistry. The emphasis was instead on other kinds of atoms containing antiprotons. As negatively charged particles, antiprotons can replace atomic electrons. With the antiproton some 200 times heavier than an electron, these exotic atoms look very different. The 'orbital' antiproton passes very close to the nucleus, giving an unprecedented close-up of the nuclear surface. Antiprotonic helium – with one or other of the two orbital electrons of ordinary helium replaced by an antiproton, showed all sorts of interesting physics effects. But antihydrogen, where the atomic nucleus would be an antiparticle, was ignored.

HOW TO MAKE AN ANTI-ATOM

At Stanford, the American theorist Stanley Brodsky had been working with Ivan Schmidt, a visitor from Chile, on quark distributions inside subnuclear particles. They re-examined the way a charged particle electromagnetically 'bounces' off an atom, and realized that these reactions could also produce electron–positron pairs. Ever since Blackett and Occhialini's pioneer study in 1932, electron–positron pairs were a copious by-product of physics experiments. The characteristic opposite whorls of electrons and positrons were seen whenever a high-energy photon of radiation transformed into matter energy. If the charged particles were antiprotons, and the velocities of a produced positron and one of the antiprotons were very close, Brodsky and Schmidt suspected the positron and antiproton might lock together to form an antihydrogen atom. The driver of a car passing a pedestrian does not have much opportunity for conversation, but two city cab drivers driving alongside each other are able to exchange information. The objective was to match the speeds of the positrons and antiprotons. By careful tuning, it might even be possible to 'breed' atoms of antimatter.

Before suggesting that someone actually tried an experiment, theorists Brodsky and Schmidt wanted to calculate how much antihydrogen could be made this way. Also at Stanford was atomic physicist Charles Munger, who had done experiments on positron production at nearby Berkeley. Munger, learning what Brodsky and Schmidt were trying to suggest, recommended going for the gas jet target approach to proton–antiproton collisions. Suspecting that the calculated antihydrogen production rate might be disappointingly low, Munger also pointed out that, in principle, it would be relatively easy to see only a few atoms of antihydrogen. Munger then took the idea to Fermilab, where it could be grafted on to the existing E760 gas jet project. This new experiment became E862. Its stated goal was – 'we propose to detect the first antihydrogen atoms'.

In July 1992, at the same time that E862 was going through the Fermilab approval pipeline, some eighty specialists met at Munich's Ludwig Maximilian University to discuss how to produce atoms of chemical antimatter. Despite the public relations appeal of antichemistry and the threat to close LEAR, CERN had no plans in this direction. However, some LEAR physicists were at the meeting. After a talk by Nobel prizewinner Wolfgang Paul and a survey of antiproton physics, the meeting got down to the real business in hand – how to synthesize antihydrogen.

Also at the Munich antihydrogen conference were three CERN specialists who knew LEAR like the back of their hands. Listening to the different possibilities for synthesizing antihydrogen, Michel Chanel, Pierre Lefèvre and Dieter Möhl remembered about the windows in LEAR's 90° bends where neutral particles could escape. They wondered if Jetset, with its gas jet target, had been making antihydrogen atoms all the time, but nobody had bothered to look. In 1993, the Jetset team cobbled together a few pieces of detector that were lying around and mounted them downstream of the detector behind the LEAR exit window. The positron would easily be stripped from the anti-atom and would register in the first sensor. The annihilation of this particle, with its characteristic back-to-back photons, would be picked up by sensitive crystals. The remaining nuclear antiprotons would be picked up in a series of semiconductor sensors. A series of sensors over a 5-metre path would also time the emerging particles. This 'stopwatch' would be able to distinguish LEAR antiprotons from other particles. For two weeks, Jetset scanned its antiprotons, covering a wide range of energies. Thirteen pairs of coincident particles were collected, of which just one showed all the right positron behaviour, with the back-to-back

FIGURE 12.3 The LEAR low-energy antiproton ring. In the foreground is a quadrant of magnets to bend the circulating antiproton beams. In the background, after the short straight section, is the modest apparatus that detected the first atoms of antimatter in 1995 (photo CERN).

photons. The team was jubilant, but a single antihydrogen candidate was not enough. The experiment was repeated the following year, but no more antihydrogen appeared.

In October 1994, 16 physicists from the 42-strong Jetset team tabled a formal proposal for a separate experiment to stake a claim to antihydrogen production. Most of the physicists were from three research institutes in Germany – Jülich (including the experiment's spokesman, Walter Oelert), the GSI Darmstadt laboratory, and Erlangen–Nürnberg University. A four-man team from Genoa University and Nuclear Physics Institute, led by Mario Macri, would look after the vital gas jet. To accomplish this meant that extra detectors had to be built and mounted both outside and inside the LEAR exit window, and the LEAR pipe itself had to be tampered with. But there was a lot of competition for LEAR's antiprotons, and the experiment had to be done as quickly as possible. To increase the rate of electron–positron production, the team proposed that the gas jet squirted a heavy gas, rather than hydrogen, into the path of the oncoming antiprotons. Their original proposal was for nitrogen or argon, but with only a few days of running open to them, they opted for the ultra-heavy xenon. The modest experiment was formally given the green light in February, 1995, and after the new detector elements were hurriedly installed, the run took place the following summer. The rest of the story has already been told in chapter 1.

Few other science discoveries have produced so much media coverage so quickly. But, ironically, most particle physicists at CERN stubbornly refused to acknowledge this public interest. The synthesis of antihydrogen merely confirmed what they already knew. The debut of antichemistry was a 'newspaper experiment', they said. At the end of 1996, just one year after the antihydrogen sensation, the executioner's axe fell and LEAR was shut down for good. In its twelve-year career it handled over a hundred million million antiprotons. However, if all these antiparticles could be brought together, they would weigh only 0.2 billionths of a gram. The total energy released by all these antiprotons annihilating was 40 Joules, enough power to light a low-wattage bulb for a second.

But all was not lost. To preserve an antimatter foothold, one of the rings of CERN's antimatter factory was modified so that antiprotons will continue to be available in Europe. Oerlert's breakthrough was only a sneak preview of antimatter. With the experiment so 'hot', antiprotons and positrons had managed to cling together as antihydrogen for only 40 billionths of a second. The real business can only begin when antihydrogen atoms can be kept still long enough for the positron to have time to make a leisurely descent down the energy ladder of antihydrogen. Only then can hydrogen and antihydrogen be compared. New experiments at CERN and at Fermilab will try to hold antimatter still and examine it closely. From this precision study of antimatter will come the most revealing indications of all – does antimatter behave like ordinary matter? How does it behave under gravity? Are the spectra of anti-atoms the same as those of atoms? Any differences would help explain why antimatter is missing in our Universe.

13 Antimatter in action

Science can be difficult to understand, and to advance science by original research is even more difficult. In his foreword to Lawrence Krauss' best-selling book *The Physics of Star Trek*, Stephen Hawking says that science fiction fulfils a serious purpose, stimulating the human imagination. 'Science fiction suggests ideas that scientists incorporate into their theories, but sometimes science turns up notions that are stranger than any science fiction', says Hawking. Antimatter is one such notion, and has continually stimulated the imaginations of science fiction writers. Antimatter is crucial to the functioning of Star Trek's USS *Enterprise* – without antimatter, there would be no Star Trek. Without spoiling the party, Krauss points out the many scientific loopholes in such antimatter-fuelled science fiction. But sometimes antimatter has fuelled the imagination of the scientists themselves.

STAR WARS

In a televised speech on 23 March 1983, US President Ronald Reagan announced the advent of a new era in US weapons development, which would 'free the world from the threat of nuclear war' by rendering nuclear missiles 'impotent and obsolete'. By the end of that year, the US Department of Defense had organized ongoing weapons research and development work under an umbrella 'Strategic Defense Initiative' (SDI) organization, headed by Air Force Lieutenant General James A. Abrahamson. The objective was an impenetrable ultra-sophisticated weapons system to knock out incoming ballistic missiles from a 10,000-strong Soviet arsenal before they reached their objective. Layered screens of sensors permanently aloft in satellites would detect the Soviet rockets from the moment they blasted off from their silos or submarines and track their progress. When the decision to intercept was taken, a screen of ingenious decoys would baffle the computers of the incoming missiles, while new 'smart' weapons would seek them out and destroy them. Even if an incoming missile launched a spray of warheads, these too would be smothered. In systems of unimaginable complexity, battlestations would orbit the Earth, in a state of permanent readiness to fire twenty-first-century laser weapons. It would be a real-life version of a Nintendo video game, and the United States boasted it would win every time.

Despite repeated White House requests to use the formal SDI label, the imagination behind the futuristic scheme soon led the media to coin the name 'Star Wars', from the 1977 George Lucas film in which the forces of good, led by Luke ('May the Force be with you') Skywalker, were pitted against the evil warriors under the command of Darth Vader. General Abrahamson and his staff soon became 'Star Warriors'. Initially some $26 billion were earmarked, and plum contracts were awarded to major contractors – TRW, Lockheed, General Electric, Rockwell, Boeing, Grumman, Hughes, Martin Marietta . . . Many of these firms set up special SDI divisions or named special vice-presidents to oversee the lucrative new outlet.

These were the balmy days of US deficit spending, huge fiscal overshoots being confidently underwritten as wave after wave of bond scrip was snapped up by an eager market. As money poured in, Star Wars' schemes became even more audacious. To the Star Wars' monitoring system, the Soviet ballistic missiles would appear ponderously slow compared to laser beams moving at the speed of light. New X-ray lasers in space-borne arrays would be triggered by nothing less than a nuclear explosion to prime the master lasing chamber, in turn feeding many smaller cylinders. The nuclear explosion would, of course, destroy the parent weapon, but by the time this happened a series of mirrors and lasing rods would have irrevocably pointed the X-ray death blasts towards their selected targets.

Once the Soviet intercontinental ballistic missiles had been boosted above the screen of the Earth's atmosphere, they would also be open to attack from powerful beams of subnuclear particles. To avoid erratic twists and turns in the magnetic field surrounding the Earth, the beams of electrically charged particles would pass through electrically neutral 'silencers' before leaving the particle gun. Complementing the array of defensive gadgets would be huge supercannon delivering tiny but smart projectiles whose on-board electronics would home in on the target. These cannon would be new 'rail guns' using electromagnetic energy instead of a chemical explosion to fire bullets at 11 kilometres per second, ten times faster than a rifle bullet. At these velocities, even plastic bullets can slice through an inch of steel plate. A fully developed rapid-fire rail gun would squirt ten bullets within a fraction of a second, but to fire them would require 2.5 gigawatts of electrical power, enough to supply a city.

Thus a major Star Wars' requirement was for reliable and flexible space-borne power sources to drive such systems. Faced with a series of decoy alerts, conventional batteries would quickly run down. Although small reactors had been lofted in satellites, major space-borne sources of nuclear power would be too expensive in terms of radiation shielding. Thus, engineers' imagination was turning to possible new power sources whose ambitious performance goals and sheer audacity was in tune with the rest of the programme.

In parallel with Star Wars, another vast US physics project was being prepared. After Carlo Rubbia had discovered the new W and Z particles in 1983 at CERN's proton–antiproton collider and walked off with a Nobel prize, some US physicists felt piqued. After absorbing a backlog of pre-Second World War discoveries, the Nobel Prize for Physics had been dominated by American researchers. In 1984, Europe suddenly reappeared on the map. Used to having it their own way, some US physicists were embarrassed and humiliated by this sudden success on the wrong side of the Atlantic. CERN was also getting ready to build a new 27-kilometre ring to collide electrons and positrons for the next generation of peaceful physics experiments. The Americans had nothing as big as that. Stung into action, the US began planning for a mighty 87-kilometre ring to accelerate and collide proton beams at the highest energies in the world. Such a large machine would have to be built where land was cheap and where there were no obstacles. First unofficially dubbed the 'Desertron', but eventually officially named the 'Superconducting Supercollider', its adherents claimed it 'reflected the universal aspirations of the US high energy physics community'. Propelled by national pride, the project pushed ahead. It would be a US scientific flagship, creating jobs and boosting the national science education programme. It would also provide trained brainpower to feed the Star Wars programme.

Common sense had suggested that the 87-kilometre machine should be built at Fermilab in Illinois, where the existing 6-kilometre ring provided a perfect injector to give the Supercollider's protons their first boost before going into orbit round the 87-kilometre ring. But common sense did not prevail. The Superconducting Supercollider soon became prey to US 'pork barrel' politics, with states vying to give it a home. Texas, a young state hungry for culture, promised lots of money, and in 1988 was chosen to host the giant new machine.

Fuelled by such giant projects, the physics atmosphere in the US in the 1980s was impassioned and frenzied. Imaginations had been let loose and money was sloshing around. It was in this heady atmosphere that the US Air Force sponsored two workshops on antimatter technology in April and October 1987 at the RAND Corporation think-tank in Santa Monica, California. RAND – short for Research and Development – had been set up in the late 1940s to advise the US Air Force on logistics policy. Subsequently, this role was broadened to advise the government as a whole on a wide range of issues, but defence matters were traditionally high on the RAND agenda. Could antimatter satisfy Star Wars' voracious energy appetite? Money did not seem to be any obstacle. The harder the problem Star Wars had to overcome, the more money was thrown at it until it finally succumbed.

On paper, antimatter is the ultimate energy source. Energy can be packaged as latent power in a lump of fuel, or stored in a device, like a spring or an electric battery. Before being used, energy has to be converted from one kind to another. When the brakes are applied on a car, the energy of the car's motion is converted to heat as the brakes rub against the wheels. Stopping a car travelling at 50 kilometres per hour produces enough heat to make a cup of tea. The ratio between kinetic energy and the heat it produces is well defined – the mechanical equivalent of heat. All forms of energy are interchangeable, and each such process has a fixed energy 'rate of exchange'. The major contribution of Einsteinian relativity was the idea that mass – the very existence of matter – is itself a form of energy. Here the rate of exchange between energy, E, and mass, m, is given by $E = mc^2$, where m is the mass and c is the speed of light. Light travels very fast – 300,000 kilometres (about seven times the earth's circumference) per second. Squared, this is a very large number, making it very difficult to convert free energy into inert mass. However, in the reverse direction, even tiny amounts of liberated mass can yield a lot of energy. In a fusion or fission bomb, only a few parts in a thousand of mass are converted into energy, but that is enough to devastate the city of Hiroshima from the air or shake Mururoa atoll from below.

All energy sources need some kind of fuel, raw material which is consumed. This consumption can either take the form of chemical burning, where the fuel combines with oxygen, or nuclear 'burning', where heavy unstable nuclei are transformed into lighter, more stable ones. In nuclear reactions, surplus mass is released as $E = mc^2$ energy. Another characteristic of energy sources is the residue left after the fuel consumption process is complete – ash, exhaust gases, or nuclear waste. Whatever the merits of the energy production process, the residue is normally unwelcome, and creates pollution problems.

However, antimatter offers the possibility of an ultimate and totally clean energy source, a dream which has stimulated the imagination of scientists ever since Dirac showed that antimatter was a natural result of the equations of relativity. When matter and antimatter annihilate, all their mass can be transformed into energy. Weight for weight, an antimatter bomb would be thousands of times more powerful than a thermonuclear weapon, a virtually unfathomable source of energy. If this release of energy could be controlled, just a few grams of antimatter could power a city for several hours. In addition, the annihilation process would be 100 per cent efficient and leaves no 'ash'.

While antimatter would provide the perfect fuel, it is also the most difficult to obtain. No fuel comes free – minerals have to be extracted from the ground, processed and transported to where they are needed. But there is no antimatter 'mine' where the stuff can be dug out of the ground. Every atom of antimatter fuel would first have to be manufactured. Supplying this energy is subject to the same $E = mc^2$ equation that the antimatter fuel would ultimately provide. No more energy can be got out than is put in. Given that antiprotons are only one of the particles produced when high-energy beams smash into a target and produce new particles, lots of $E = mc^2$ energy is lost as unwanted particles of ordinary matter. The high-energy beam cannot be 'tuned' to produce only antiparticles – a million protons have to be accelerated and smashed into a target to produce on average just one useful antiproton. All the antiprotons produced at CERN during one year would supply enough energy to light a 100 watt electric bulb for three seconds! In terms of the energy put in to produce high-energy proton beams and store them, the efficiency of the antimatter energy production process would be 0.00000001 per cent. Even the steam-engine is millions of times more efficient!

There is probably another catch to using antimatter energy. If a drop of water is put on a boiling-hot surface, it evaporates quickly, making a loud hissing noise. Increasing the temperature of the surface before putting a second drop of water on it causes the second drop to fiercely crackle, evaporating almost explosively. However, put the drop on a red-hot surface and its stays quietly in position, possibly oscillating slowly from one side to the other as it slowly evaporates. This apparently paradoxical behaviour, called the Leidenfrost effect after the German physician who discovered it in the nineteenth century, happens because a thin layer of steam insulates the drop of water from the extreme temperature beneath. Lower the temperature of the plate and the steam layer becomes thinner, decreasing the insulation between the water drop and the hot plate. At a certain temperature the steam ceases to insulate the water drop any more and the drop explodes. The moral of this story is that, with a powerful potential source of energy, an initial layer of reaction product will insulate the remainder of the energy source and prevent it from coming into play. With antimatter, the radiation produced by the first contact between matter and anti-matter would act as a cushion, shielding the rest of the antimatter.

But, in the heady days of Star Wars, people did not want to listen to such pessimistic arguments. If enough money were available, an anti-matter battery, in principle, could be put in orbit to drive a Star Wars' battlestation array. The United States Air Force wanted ideas, not warnings. No holds were barred, and the sky was the limit. A direct outcome was the ARIES (Applied Research In Energy Storage) project, set up in 1986 at the Air Force Rocket Propulsion Laboratories at Edwards Air Force Base in California. The following year, about eighty scientists participated in the USAF-sponsored RAND Corporation workshops on antimatter technology. The resulting report recom-mended construction of a US antiproton source. RAND physicist and conference organizer Bruno Augenstein said the workshop findings 'indicate that we are on the threshold of important advances both in the basic science of antimatter and its practical applications. The work-shop participants urge the nation to get on with supporting it.'

The report called for initial design and feasibility experiments to pave the way for full-scale antimatter propulsion systems. The enor-mous energy released by matter–antimatter annihilation would pro-vide new propulsion systems, said the report, adding, however, that this would have to await the availability of milligram quantities of antimatter. But, whatever the application, handling antimatter would require special traps, which would be used to transport the precious antimatter from the production centre to the laboratory which would undertake the next stage of the research. The development of these antimatter storage jars was seen as a 'critical enabling tool'.

As these projects were being orchestrated, the world political cli-mate suddenly changed beyond recognition. The Berlin Wall fell in 1989, and within two years the Soviet Union itself had been dismembered.

FIGURE 13.1 The 87-kilometre US Superconducting Supercollider (SSC) near Waxahachie in Texas was to have been the world's largest scientific instrument. This 1991 aerial photo shows an experimental hall, the size of a football field, built above the tunnel being built to house the Supercollider. The machine was abruptly cancelled in 1993, leaving 23 kilometres of empty tunnel (photo SSC).

Almost overnight in historical terms, half a century of Cold War evaporated. Star Wars became irrelevant and a major part of the motivation which had underlined the massive research push disappeared. For the Superconducting Supercollider, the Second World War nuclear bomb-makers, the traditional godfathers of particle physics, were either dead or no longer influential. On the financial front, a huge accumulated US budget deficit could no longer be swept under the financial carpet and the fiscal brakes had to be firmly applied. Squeezed by this pincer movement, the Star Wars programme crashed to a standstill and the Superconducting Supercollider was halted in 1993 with 23 kilometres of tunnel near Waxahachie, south of Dallas, lying empty. Many scientific careers looked to be in ruins, but sophisticated mathematical skills were quickly snapped up by astute investment-fund managers looking for specialists who could make computer predictions to beat the market. While few people would have predicted in 1985 that the Berlin Wall would fall before the end of the decade, perhaps Star Wars and the other 1980s US scientific razzmatazz helped achieve an objective without anybody having to press a button. The threat of US scientific and technological know-how was certainly influential in undermining Soviet confidence and hastening the downfall of an outmoded system.

PICTURES FROM ANTIPARTICLES

With Star Wars on the scrap-heap, antimatter physics was put on a slow but more realistic track. Plans to develop antimatter engines were put aside, and antiprotons and positrons became stock-in-trade for routine physics experiments. With their help, new particles were discovered, and the properties of known ones carefully documented. But, in the meantime, antiparticles had been tamed for another use, this time in medicine and materials science, making detailed radiographs to make the invisible visible.

The history of imaging began in the afternoon of 8 November 1895, when Wilhelm Konrad Röntgen, Professor of Physics at the University of Würzburg, Germany, was preparing an experiment to pass a high voltage through a vacuum tube. Röntgen had heard other researchers report how the high-voltage discharge – cathode rays – could pass through different substances, but wanted to see for himself. Covering his cathode ray tube with thick black paper, he darkened the room and switched on the voltage. Suddenly he was surprised to see a fluorescent screen on the other side of the room begin to glow. Some sort of radiation was passing through the black paper and causing the screen to shine. Röntgen tried putting other obstacles between the cathode ray tube and the screen, and was startled to see on the screen an image of the bones in his hand. The radiation, which he called X-rays, was blocked by bones, but shone easily through the surrounding flesh. The rays themselves were invisible, but could affect a sensitive fluorescent screen or a photographic plate and produce an image.

Within a year, Röntgen's chance discovery was being used in medicine to view bone fractures and in dentistry to search for cavities in teeth. For the next seventy years, X-ray technology remained basically as Röntgen had discovered it, a cathode ray tube to produce the X-rays and a photographic plate to record the image. In 1972, a new technique was unveiled – Computer-Aided Tomography (CAT), from the Greek 'tomos', slice. It revolutionized the whole field of medical imaging. Instead of recording a flat image, like a conventional X-ray photograph, a CAT camera instead rotates around the patient. From the recorded information, a computer reconstructs an image of a horizontal section of the patient's body. Each rotation gives a reconstruction of a different horizontal slice, and a three-dimensional picture is built up from the successive slices.

As well as X-rays, medicine had also discovered the usefulness of radioactive tracers, isotopes that collected in certain organs and whose radiation gave a sharp image when viewed by a suitable detector or camera. These techniques – 'nuclear medicine' – complemented the information available via X-rays. The radiation emitted by tracers gave accurate images of the brain and the heart.

Soon after the discovery of the positron in 1932, Frederic Joliot and Irène Curie discovered new radioactive materials that emitted positrons rather than electrons. Normally, radioactive decay produces electrons as unstable neutrons decay into protons. The positive charge of the resulting proton is balanced by the negative charge of the ejected electron. However, in a few cases, a nucleus feels happier if it can acquire an extra neutron – a nuclear proton therefore transforms into a neutron, emitting a positron. These isotopes remained a physics curiosity until the early 1950s, when medical researchers realized that positron emitters could offer interesting new possibilities in radiography.

Positron-emitting tracers work in a different way to other forms of radioactive labelling. Instead of the emitting particles being picked up by a waiting detector, antiparticles quickly annihilate with atomic electrons *in situ*. When this happens, the annihilation energy – the combined mass of the electron and the positron – is converted into two bursts of high-energy radiation – gamma rays. These two gamma rays, moreover, have to emerge back-to-back to balance the overall momentum. If these pairs of gamma rays could be picked up, they would point back to where the annihilation happened. A picture of the tracer site could be built up point by point, even without the need for tomography.

With the image quicker to produce, this technique would require less tracer and a correspondingly reduced radiation dose for the patient. As if to underline the potential usefulness of the method, Nature had thoughtfully provided oxygen, fluorine, nitrogen and carbon with positron-emitting isotopes, whose lifetimes are long enough to get the tracer into the patient, but with manageable exposure times.

The major obstacle was the development of a suitable camera to record the gamma rays, and it took almost thirty years of development work, mainly in perfecting gamma-ray detectors for physics laboratory experiments, before the first commercial instruments became available. With the necessary technology available, Positron Emission Tomography (PET) became a medical standard and was also adopted for remote imaging in materials science, where, for example, it can follow the course of oil through an engine. The main difficulty is the availability of the unstable positron-emitting isotopes, which have to be produced nearby, but there are now about 150 centres worldwide equipped for PET scans. Applications include examination of tumours, localization of seizure foci in epileptic patients and investigation of other neurological problems, and assessment of tissue viability prior to cardiac surgery. PET scans provide incisive information and frequently avoid unneccesary 'blind' surgical exploration. After medical treatment, biochemical results can be seen with PET scans even before any biological signs become visible.

PET enables the take-up of a single biochemical to be followed in a living subject. Before the advent of PET, this study could only be done by injecting animals which were then sacrificed after a certain time, frozen and sectioned. Repeating the process over a time-scale enabled researchers to follow how the substance was taken up, eventually

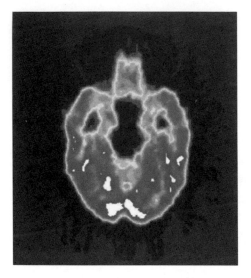

FIGURE 13.2 A positron-emission tomograph (PET) picture of a human brain. After an injection of a positron-rich tracer, the positrons 'go to the head' of the patient and the subsequent annihilations of the positrons with electrons in atoms in the brain reveal the brain's structure. PET can help reveal the brain centres responsible for epilepsy, and diagnose degenerative disorders such as Alzheimer's and Parkinson's diseases (photo Adrian McKemey, Brunel University and MRC, UK).

obtaining an overall biological 'model' which was of limited use subjectively.

Powerful particle beams cause atomic havoc when they hit a target. But this destructive power can be put to good use. Beams of particles of all kinds are increasingly being used as precision 'scalpels' to irradiate and destroy tumours deep inside the body, or in sites, like the head or the eye, which are difficult to access by conventional surgery. Carefully designed beams can pass through surrounding tissue and attack only the localized tumour. Using beams of selected nuclei, such as carbon, the nuclear reactions inside the body produce their own positron-emitting isotopes. Using a PET camera, the irradiation produces it own image of how it is destroying the tumour.

But it is in the study of the brain where PET is making a special impact. In neurology, PET can help reveal the brain centres responsible for epilepsy, and diagnose degenerative disorders such as Alzheimer's and Parkinson's diseases. In healthy brains, the biochemical effect of different compounds enables PET to pick up the tiny changes in blood flow when different centres are activated by external stimuli. Centres for language, vision, motion, colour, memory and even pain have been identified. Antiparticles have helped extend our understanding of consciousness itself.

Ingenious techniques have also been developed to use low-energy positrons and their annihilations with matter to study the structure of surfaces and for depth profiling. Several laboratories in Europe, the US and Japan have developed high-intensity positron guns for this work.

14 Antimatter of the utmost gravity

Beauty has symmetry. The most beautiful vistas, whether they be faces or buildings, are those which are the most symmetric, the most perfect. Any visible asymmetry is an immediate flaw on the face of beauty. One of the world's great architectural triumphs is the Taj Mahal at Agra, India. Constructed by an army of some 20,000 workers over thirty years, the mighty edifice was built by the Mogul emperor Shah Jahan as a monument to his beloved wife Mumtaz Mahal who died after giving birth to her fourteenth child in 1631.

Seen from afar, the splendour of the Taj Mahal is breathtaking. No visitor can fail to be impressed by its grandeur and beauty. Every dome and every tower is faithfully balanced. But the mighty monument has another dimension, just as cleverly designed and equally impressive. Walking towards the shrine, at close quarters the spectator is dwarfed by the magnitude of the construction and can no longer appreciate its symmetry. As if to compensate for this loss, the spectator becomes instead captivated by the glorious mass of intricate inlay work decorating the vast marble façades. Both the large-scale and small-scale aspects of the master design are masterpieces of beauty. It has large-scale and small-scale symmetries, the latter a mass of detail complementing the former. But without one, the other could not exist. Each is equally impressive and equally important to the success of the monument.

The Taj Mahal is perfectly symmetric, every detail balanced on the left and the right. A mirror-image of the main Taj building would be virtually indistinguishable from the real thing. The only clue would come from verses of the Koran elaborately inscribed on the lower façades, but here there was no alternative – writing, with its strict commitment to right-to-left (in the case of Arabic), cannot be counterbalanced. In front,

FIGURE 14.1 Perfect symmetry – the Taj Mahal (photo G. Fraser).

a long pool reflects the building vertically, adding to the symmetry. As well as the main shrine, the surrounding complex too is symmetric, the landscaped gardens providing a perfectly balanced foreground. On each side of the shrine are two smaller and identical buildings, one facing the shrine from the east, the other from the west. The one to the west is a mosque. One mosque is enough for the faithful, its mirror-image serving no purpose other than to visually counterbalance the mosque in the west.

The perfect but nevertheless superficial symmetry of one of the world's great buildings mirrors that of our Universe as a whole, which has symmetry on different levels. These symmetries were dictated when the Universe was created some 15,000 million years ago as a quantum pin-point of hot, dense matter erupted into a fireball of energy – the Big Bang. All primordial matter in the Universe came from this single cataclysm, and, to balance its books, the explosion should have created equal amounts of matter and antimatter. The Big Bang should have been matter–antimatter symmetric. But the visible Universe around us shows little sign of this primordial antimatter. Like the grand plan of the Taj Mahal, perhaps antimatter is there to ensure an overall superficial symmetry which we can no longer see from our viewpoint.

THE MISSING ANTIMATTER

The idea of a Big Bang origin of the Universe was put together in the 1920s and 1930s, but initially not all scientists took notice. Many believed in a 'Steady State' picture, in which the Universe always has existed and will continue to exist, with new matter being created all the time, like water from a slowly dripping tap. Paul Dirac, the spiritual father of antimatter, probably did not yet know very much about what would ultimately become the Big Bang picture when he gave his Nobel lecture on 12 December 1933 and suggested that the Universe could contain both matter and antimatter without us knowing (see chapter 4).

If Dirac were right, the whole Universe should be a uniform mix of matter and antimatter. In some places there might be more matter than antimatter, or vice versa, but overall the two halves of the Universe should balance. Where is this antimatter? Certainly there is no accumulated antimatter on Earth, nor even in the Solar System, otherwise we would have seen the result of its violent encounters with ordinary matter. At the centre of the solar system, the Sun steadily burns hydrogen into helium, giving off a steady stream of particles, the 'solar wind'. If this wind were to meet any antimatter in its path, there would be brilliant flashes of matter–antimatter annihilation.

As well as luminous stars and galaxies, the Universe is also filled with dark cosmic dust. Some of this material is virgin and has yet to find a stellar home, some comes from old stars that have exploded and flung their ashes into the depths of space. A star is fuelled by thermonuclear fusion – as light nuclei are cooked into heavier ones, energy is released and the star shines. (Some positrons do exist transiently in the depths of the Sun's interior, and presumably in the depths of other stars too, where they are formed in the thermonuclear fusion of protons – the first step towards solar energy; see chapter 5.) As each thermonuclear process is complete, a new process takes over until eventually all thermonuclear processes have been exploited. With its nuclear fuel exhausted, the star can no longer resist the crush of its own gravity and begins to crumple. However, this gravitational compression eventually reaches a limit when the stellar material cannot be compressed any further. The superdense remnant of the star then springs back like a rubber ball that has been squeezed and explodes as a mighty 'supernova', flinging stellar debris far out into space.

Whatever its origin, primordial matter or stellar ash, cosmic dust is never uniform, and wisps of it gradually clump together under the pull of gravity, the denser regions accumulating more dust. Ultimately enough matter collects together for a new star to light up. On Earth, this dust is seen as cosmic rays, electrically charged subatomic particles accelerated to very high energies as they whirl through the magnetic fields of the cosmos. After traversing light-years of empty space, some extra-terrestrial particles smash into the relatively dense upper layers of the Earth's atmosphere, creating cascades of new particles which shower down to the Earth's surface. Recording these cosmic cascades in detectors, physicists found new types of particle that had never been seen on Earth – examples being the positron and the kaon.

If the Universe contains antimatter, then cosmic rays should also contain antiparticles. Light antiparticles, such as positrons, are common in cosmic rays. However, such light antiparticles are usually from particle–antiparticle pairs produced from the energy dispersed as primary cosmic ray particles collide with atmospheric gas or interstellar dust. They are not necessarily primordial cosmic antimatter. 'Fountains' of positrons have been seen by satellite-borne detectors peering into the centre of our Galaxy, but these can be explained by violent cosmic processes spitting out huge amounts of radiation that can easily make electron–positron pairs. There are no fountains of any other sort of antiparticle.

Any antimatter stars would contain antinuclei, mirrors of the nuclei we know but built from antiprotons and antineutrons. When such stars died in supernova explosions, their stew of antinuclei would have been flung out into space. But the cosmic rays arriving at the Earth's surface or even in the upper atmosphere have revealed no sign of antimatter heavier than antiprotons. Where has all the Big Bang antimatter gone?

If we cannot see any antimatter, perhaps matter and antimatter are separated into distinct domains. The Universe we know is a matter domain. Maybe somewhere else there is a corresponding antimatter domain – an antidomain. These mirror Universes could have lost contact with each other and gone their separate ways. But, even so, they should have been in contact just after they were created in the Big Bang. Wherever and whenever the boundaries of the domain and antidomain briefly touched, pieces of matter and antimatter would have mutually annihilated to give powerful bursts of radiant energy – what physicists call gamma rays. As the Universe subsequently cooled down, these gamma rays would have cooled down too and produced a dim but uniform cosmic gamma-ray signal all over the sky. Knowing the energy that would have been liberated in such a primordial matter–antimatter encounter, physicists can estimate what such a signal should look like 15,000 million years later.

In 1991, the Space Shuttle Atlantis placed into orbit a new eye in the sky, the Gamma Ray Observatory (GRO). Physicists were able to view cosmic gamma rays beyond the curtain of the Earth's atmosphere. Gamma-ray astronomy had in fact been born in 1967, when the US 'spy' satellite Vela was deployed to look for the tell-tale gamma-ray bursts of Soviet nuclear explosions. Vela saw gamma-ray bursts, but, instead of coming from Earth, they came from outer space. From its vantage-point, GRO clearly saw these bursts against a faint but uniform gamma-ray backdrop. The bursts are more interesting than the faint backdrop, but physicists saw that this backdrop is feebler than what would have resulted from primordial large-scale matter–antimatter annihilation. Today's gamma-ray background shows no sign of matter–antimatter annihilation processes ever having taken place on a large scale.

MICROWAVE PATTERNS

Perhaps the matter and antimatter domains were never in contact with each other? After they were formed in the Big Bang, matter and antimatter immediately went their own separate ways, blown apart by the force of the primeval explosion. In this case, the separate matter and antimatter domains in the Universe would be separated by immense voids. But such a patchy Universe would have a special signature.

Just a tiny fraction of a second after the Big Bang, the cosmic soup cooled sufficiently for quarks to stick together as particles like protons and neutrons, and for antiquarks to form antiprotons and antineutrons. After another 100 seconds or so, these nuclear particles cooled sufficiently to form light nuclei like helium and antihelium.

300,000 years later, two events took place (300,000 years sounds a long time, but in fact is the same fraction of the 15,000 million years of the Universe's existence as a few minutes are of one year). The cosmic broth became cool enough for electrons to stick to protons or other light nuclei – the first atoms were formed. In antimatter regions, positrons would have stuck to light antinuclei to give atoms of chemical antimatter. With radiation no longer absorbed by the cosmic material, the Universe suddenly became transparent and 'there was light'. A blinding cosmic flash signalled the advent of atomic matter. This flash continued to cool as the Universe expanded, eventually becoming a faint microwave glimmer. This glimmer was detected by Arno Penzias and Robert Wilson at Bell Laboratories, New Jersey, in 1965. Developing sensitive antennae to pick up signals from the new communications satellites, Penzias and Wilson had been plagued by an irritating hiss which at first they thought was interference. They tried valiantly to make the hiss disappear, but could not, and finally concluded it was a signal from deep in the Universe, a characteristic of space itself. This Cosmic Background Radiation is the last rumble of the Big Bang. Penzias and Wilson's discovery showed that the Big Bang theory was on the right track, and alternative ideas such as the Steady State theory fell by the wayside.

The Cosmic Background Radiation fills all space and is extremely smooth. But it cannot be perfectly smooth, as the Universe that created it was also not perfectly smooth. Physicists like Stephen Hawking pointed out that the Cosmic Background Radiation should reflect the structure of the Universe as it was some fifteen billion years ago. These irregularities of matter (and antimatter) were the gravitational seeds which grew into the galactic structures we see today.

To look for these seeds of gravitational evolution, in the early 1980s a team led by John Mather of NASA proposed the Cosmic Background Explorer (COBE) satellite for launch aboard the Space Shuttle. The 1986 Challenger disaster temporarily held up this programme, and COBE was hurriedly redesigned to fit a conventional rocket, going into orbit in 1990. In 1992, COBE duly found ripples in the Cosmic Background Radiation. This tiny temperature shimmer, just 30 millionths of a degree, provided the first map of the proto-Universe. This tiny stencil guided gravity's brush as it continued to paint the evolving Universe. COBE's findings subsequently have been confirmed and extended by other instruments.

Cosmologists now have to understand how such tiny seeds evolved into the galaxies we now see. But one thing is already clear. If the initial Universe had contained widely spaced clusters of matter and antimatter, these would have left their characteristic imprint on the Cosmic Background Radiation. The tiny ripples seen by COBE and other detectors are not compatible with separate domains of matter and antimatter that went their own ways immediately after the Big Bang. All primeval antimatter was thus exposed to contact, and therefore annihilation, with matter. Annihilation would have been unavoidable, but the resultant cosmic radiation shows no signs of it. The Universe we can see looks to have been eternally free of nuclear antimatter.

GRAVITY CAN PUSH AS WELL AS PULL

Why has antimatter become irrelevant? In chapter 8 we saw how the subtle asymmetries at work at the quark level could have helped erode antimatter from the face of creation. Another culprit could be gravity. Together these mechanisms could have ensured that antimatter symmetry is a largely empty concept for our Universe, just as the Taj Mahal's 'mirror' mosque serves no purpose other than to maintain the symmetry of the Grand Design.

The evolution of the whole Universe has been and still is controlled by the all-pervading force of gravity, and gravity will ultimately seal its fate. The force of gravity we know best is a pull. In the seventeenth century, Isaac Newton realized that all masses exert a gravitational pull on each other, acting as though their masses are concentrated at their 'centre of gravity'. Under this remorseless attraction, apples fall from trees while stars are locked in their eternal orbits. Newton's realization that these two very different effects were manifestations of the same force was a major breakthrough in understanding. Albert Einstein, in his 1916 theory of general relativity, went one giant step further. Einstein explained that matter deforms the space and time around it, like a large weight placed in the centre of a rubber sheet. Smaller weights placed elsewhere on the sheet make their own dents, but fall towards the large weight because of the way the sheet sags. In the same way, said Einstein, masses follow the path of least resistance through a space and time deformed by massive objects. The mystical 'force' of gravity is simply the geometry of space and time. With no matter around, space and time would be flat and a hypothetical 'test' particle would not roll anywhere – there would be no gravity.

Antigravity, the gravitational force between masses of antimatter, should also be an attraction, but this can only be verified by an anti-Galileo collecting some antimatter and doing experiments at the leaning anti-tower of anti-Pisa. New experiments at antiproton sources will have to content themselves by trying to measure how antimatter behaves under ordinary terrestrial gravity (see chapter 11). If antimatter 'falls up', this would at first seem unfamiliar. But it would not be the first time that gravity has been a repulsive force, pushing bodies violently away from each other. Gravity has a second face, less familiar than its everyday pull, but no less important for the Universe as a whole. Gravity can also repel, and without this repulsion, the Universe would not exist.

Space and time are never empty. Even before the creation of the Universe the dimensions of existence were filled with quantum shimmer as transient particle–antiparticle pairs briefly lit up, temporarily defying the absence of enough energy to support their existence. Most of this shimmer faded without leaving a trace, but there was at least one exceptional case – the Big Bang. If a certain probability of finding a transient quantum pin-point is plugged into Einstein's equations of general relativity, the equations reveal that the quantum bubble expands faster even than the speed of light, doubling its size in just 10^{-34} of a second. Called 'inflation' by cosmologists (see chapter 8), this very different face of gravity amplified a transient quantum pin-point of hot, dense matter and antimatter into a viable football-sized 'Universe'. Having been forced into existence, the matter and antimatter had to resolve their differences, and the outcome then began to see the second, gentler, face of gravity. Matter tried to pull itself together and halt this initial expansion. This tug-of-war between the initial expansion phase of gravity and the subsequent pull of aggregate matter has been continuing ever since.

If the Universe contains enough matter, the gravitational attraction between this matter will ultimately stop the Universe expanding. The pull of gravity will have finally overcome its initial push and the Universe will begin instead to contract, ultimately collapsing in a 'Big Crunch'. If it does not contain enough matter, the Universe will continue expanding for ever under the Big Bang's repulsive gravity. With our stocktaking of matter in the Universe still nowhere near complete, cosmologists can only speculate which of these two scenarios is correct.

The ultimate Big Crunch, if it happens, will destroy the Universe. But less significant crunches due to the remorseless pull of gravity are going on all the time. The 'critical' velocity a rocket needs to attain before it can escape the grip of gravity depends on the mass of the star or planet it is trying to escape from. The velocity needed to escape from the Earth is about 11 kilometres per second, while to escape from the Moon only about 2.4 kilometres per second is needed. For much heavier stars, this escape velocity eventually reaches that of light, so that not even light itself can escape from the star. Such heavy stars are condemned by gravity to be an eternal 'black hole', a flaw on the face of the Universe swallowing whatever comes near. Black holes show no sign of what they once were – a black hole caused by the collapse of an antimatter star would look just the same as any other black hole. The antimatter of the Universe could have been locked away in black holes.

THE EXPANDING UNIVERSE

The eternal struggle between the attractive and repulsive aspects of gravity can be seen in the way the Universe expands. Only a hundred years ago, astronomers believed that our galaxy, the Milky Way, contained all the stars in the Universe. The Universe could be larger, they said, but this surrounding void was empty and of no interest. In a twentieth-century Copernican revolution, the American astronomer Edwin Hubble discovered that the solid material of the Universe extends much further than our own galaxy, and that these distant galaxies appear to be rushing away from each other. The further away a galaxy is, the longer its light has taken to reach us and the younger the image of the galaxy. These distant galaxies appeared to recede faster than those nearer to us (and which therefore appear older) – the famous 'Hubble expansion'. The dim light from these distant galaxies, billions of light-years away, was emitted when the Universe was young and the inflation aftermath of the Big Bang had not yet been fully compromised by the pull between the matter it contains.

Measuring the rate at which these distant galaxies recede enables cosmologists to estimate the age of the Universe. The further away a galaxy is, the faster it appears to recede. The faster this apparent expansion, the more time has passed since the Big Bang. These measurements are difficult because the remote galaxies require independent distance measurements. Thus there have been periodic reports that the Universe is younger than its oldest stars – the 'old wine in new bottles' dilemma. These problems have been resolved in recent years by the wonderful sharp images given by the Hubble Space Telescope from its extra-terrestrial vantage-point.

A new slant on the way the Universe evolves has come from ground-based telescopes scanning for the galactic explosions of supernovae. Certain supernovae are always produced in the same way and all the resulting explosions should be equally powerful. All such supernovae should therefore be equally bright, and comparing their apparent strength as viewed from Earth gives a reliable measure of their relative distance. In recent years, telescopes have collected supernova data and, by comparing the signals, astronomers now see that a Hubble expansion – that inexorably slows due to the mutual gravitational attraction of objects – is not the whole story. The expansion of the Universe now appears to be gently speeding up rather than slowing down – a powerful gravitational repulsion is still at work alongside the more familiar gravitational attraction. As the Universe expands and its matter is pushed further apart, the pull of gravity falls off and matter starts to feel again the push of gravity. The everyday image of gravity as an attractive force that holds planets in their orbits and makes things fall to the ground is only a 'local' picture, valid for distances far less than those travelled by a light ray since the Big Bang.

To allow for these effects, bold theorists have constructed new pictures. In addition to the conventional attraction between all masses (whether they be matter or antimatter), this introduces a new component of gravity in which matter has a sign, like electric charge. In the same way that like charges repel and unlike charges attract, the new face of gravity is repulsive between matter and matter, or antimatter and antimatter, but is attractive between matter and antimatter. As well as the delicate matter–antimatter asymmetries proposed by Sakharov, such unfamiliar gravitational effects also helped mould the Universe and seal the fate of antimatter.

THE SEARCH FOR COSMIC ANTIMATTER
When it blasted off from NASA's Kennedy Space Center on 2 June 1998, the Space Shuttle 'Discovery' had on board the 2-ton Alpha Magnetic

FIGURE 14.2 June 1998 – the Space Shuttle Discovery, with payload doors open, viewed from the Russian Mir station prior to Shuttle–Mir docking. As well as the large rectangular container with equipment for Mir, Discovery also carried a smaller module (aft) with the Alpha Magnetic Spectrometer (AMS). This was a precursor study before deploying AMS on the International Space Station, where its mission will be to look for nuclear cosmic antimatter (photo NASA).

Spectrometer (AMS), the first major particle physics experiment to go into orbit around the earth. This ten-day precursor study provided valuable experience before redeploying AMS on one of the great scientific projects for the twenty-first century, the International Space Station. The objective of AMS will be to look for nuclear cosmic antimatter. From its orbital vantage-point, the high technology AMS detector will carefully monitor the composition of cosmic rays high above the protective screen of the Earth's atmosphere. If it does find nuclear cosmic antimatter, it will help dispel the conundrum of a creation which apparently has contrived to hide half of itself. Although therefore welcome, such a positive result will nevertheless clash with the negative evidence for antimatter among all the cosmic signals trawled so far from the depths of the visible Universe. Our understanding of cosmology and the origin of the Universe would require a major rethink, a Copernican revolution for the twenty-first century.

Names index

Subject index